Foundations and Strategies for Medical Device Design

VIKKI HAZELWOOD, Ph.D., FNAI

New York   Chicago   San Francisco
Athens   London   Madrid
Mexico City   Milan   New Delhi
Singapore   Sydney   Toronto

Library of Congress Control Number: 2021936109

Foundations and Strategies for Medical Device Design

1 2 3 4 5 6 7 8 9   CCD   26 25 24 23 22 21

ISBN   978-1-260-45695-0
MHID   1-260-45695-1

This book is printed on acid-free paper.

Sponsoring Editor	**Copy Editor**
Robin Najar	James Madru
Editing Supervisor	**Proofreader**
Stephen M. Smith	Claire Splan
Production Supervisor	**Indexer**
Pamela A. Pelton	Claire Splan
Acquisitions Coordinator	**Art Director, Cover**
Elizabeth M. Houde	Jeff Weeks
Project Manager	**Composition**
Patricia Wallenburg, TypeWriting	TypeWriting

Thank you to my mother, Barbara, father, Al,
and brother, Alf, for always supporting me.

About the Author

Born in a time when general practitioners still made house calls, tooth cavities were drilled without novocaine, kids routinely had tonsils removed, and women didn't seek STEM careers, Dr. Vikki Hazelwood witnessed the rapidly advancing field of medicine as she grew up.

Today she is an academic with extensive medical industry experience, having held executive positions in sales and business development for medical biomaterials companies focused on the introduction of new devices for orthopedic and cardiac surgery as well as for drug delivery. In these roles, she worked closely with surgeons and area hospitals in a clinical setting and collaborated on numerous development projects with medical device manufacturers.

Originally trained as a chemical engineer, Dr. Hazelwood had a successful early career designing engineered equipment for chemical and pharmaceutical manufacturers. After earning an advanced degree in biomedical engineering, she moved into the field of medical devices. Despite all her years of engineering training and experience, she entered a world full of new rules for medical design. Every day posed new lessons that you don't always learn in school or in a book, until now.

Dr. Hazelwood is co-founder of the biomedical engineering program at Stevens Institute of Technology, Hoboken, New Jersey, where she leads the senior design curriculum that formally indoctrinates all of her students in the medical device design process from invention to prototyping and clinical testing. As professor in the Department of Biomedical Engineering, she established the Lab for Translation Research at Stevens, and SPOC, a medical device company. Her research focuses on the development of efficacious medical devices and physiologic methods that improve public health. She and her students have created innovations for use in emergency medicine, pain management, and physical rehabilitation.

Dr. Hazelwood has held clinical research and instructor appointments in the Emergency Department at Hackensack Meridian Health and its related medical schools. She's been PI (principal investigator) or co-PI for more than a dozen clinical trials aimed at bringing new technologies to practice.

Based upon her own experiences, she presents the field of medical device design. She offers the unique perspective of an engineer who may have significant design expertise but is relatively new to the field of medical devices. She shares what she has effectively developed for her students and for the younger faculty whom she has mentored.

Dr. Hazelwood holds a Ph.D. in biomedical engineering. She received the Advancement of Invention Award from the New Jersey Inventors Hall of Fame and multiple awards for teaching excellence at Stevens. She holds four U.S. medical device patents with her student co-inventors. Three patents have been licensed to three different companies including one start-up company that she founded, and her students remained engaged in these companies after graduation. She has written 25 articles, 1 book, and 2 book chapters related to innovation-education and advancing medical technology. Dr. Hazelwood is a Fellow of the National Academy of Inventors.

Contents

CHAPTER 16
The Landscape for Medical Devices in the Twenty-First Century . . 195

Acknowledgments

I am fortunate to have several colleagues and friends who have supported this work. I thank all who allowed me to discuss ideas and challenges throughout this project. I would especially like to thank a few who offered a lot of help.

Luckily, a bright and enthusiastic Ph.D. candidate, Emily Ashbolt, joined me when my draft manuscript needed to be constructed into a textbook. Emily contributed many hours of editing support and production preparation. She also provided the invaluable perspective of a student—cheerfully. I expect that Emily will write her own book someday soon, and when she does, I am certain it will be very, very good.

Imagine being at a backyard barbeque and asking your friends, "Hey, do you want to read the textbook I'm writing?" Most would probably shy away. I offer sincere thanks to a special friend who said, "Sure!" Thank you Karen Haas for your thoughtful critiques and for your encouragement.

I would also like to express my appreciation to my young friend Gal Bejerano for reading and offering comments on the final draft.

Lastly, I offer gratitude to my mentor, Dr. Arthur Ritter, and to all of my wonderful students who inspired me to write this book.

Message to Students

This book will help you to become a better medical device designer. If you have not yet begun your career, it will provide you with a foundation to develop successful medical device design strategies on your own. If you are already working in the field, it will give you a deeper appreciation for current practices.

It tells about events that shaped medical device regulations and standards. It reviews the past century of medical device achievements and catastrophes. It describes the ramifications of those events and how guidelines and rules for medical device design were formulated or changed.

It tells how we've failed and what we need to do better. It reviews current technologies, and it offers goals for you to consider when you design a medical device. It suggests ideas about where you can contribute to advance a design and improve health outcomes. It will help you understand why we do it this way today.

This book is not intended to be a comprehensive design manual. Rather, it offers a thorough picture of the state of the art of medical device design. By studying these lessons, you will build confidence in your ability to navigate medical device design strategies and the complex scientific, clinical, and commercial issues that are inextricably linked within them.

The events and examples in this book were selected because they represent major medical challenges, past and present. In its entirety, this book reviews the top causes of death, the most expensive diseases and illnesses, and current issues with common medical devices. During your career, you will inevitably work to address one or more of these challenges.

Each lesson summarizes at least one critical element of design strategy for medical devices. There are many stories that describe a disease and the clinical application of a medical device. You may be more familiar with some fields than others. If needed, you or your instructor can prepare for these lessons by reviewing the etiology, pathology, or business case related to the story.

The material presented in this book includes highlights from several hundred references. Most of the references are publicly available. If you want to know more about a topic, you are encouraged to review the references in their entirety. If you study these lessons well, I am sure that you will become a confident, efficient, and thoughtful medical device designer who is certain to improve the quality of someone's life.

Message to Instructors

I hope you enjoy teaching these lessons as much as I enjoyed creating them. This book will help you teach about the foundations of medical device design practice and offer insight into the rationale for strategies applied in those methods.

The examples and anecdotes were designed to incorporate the top causes of death and the most expensive diseases and illnesses as medical devices pertain to them. Therefore, this book aims to expose students to common and broadly relevant current issues. By providing lessons in this context, I expect students to begin to develop their own ability to identify opportunities to improve medical devices in order to address public health needs.

"Learning Objectives" are provided at the beginning of each chapter. "Study Questions" are provided at the end of each chapter to help guide independent student review or discussions based on the information presented in the chapter. Additionally, "Thought Questions" are provided to help students acquire a broader background in the topic of the chapter. To answer these questions, further research may be required.

"New Terms" are listed at the beginning of each chapter. These terms are set in boldface type when they are introduced. It may be helpful to review the new terms with the students before they read the chapter or during class discussion after they read the chapter.

Within many of the chapters, there are occasional sections identified as "Important Design Knowledge." These sections provide information related to common design issues found in practice. It is very likely that students planning a career in the medical device industry will encounter similar situations. The notes in these sections are worthy of emphasis.

The material presented in this book includes highlights from references, most of which are publicly available. Every effort was made to use references that are easily accessible to you and your students. The references selected were ones that might make a good supplemental reading assignment for an upper-level student. Collectively, the references include a wealth of images that are valuable for educational use.

Chapters 1–5 recount pivotal events in the past century that influenced medical device design practices. These chapters aim to lead students to appreciate the rationale for current medical device design strategies. They also demonstrate the multidisciplinary nature of biomedical engineering and provide perspective on the rapid and recent development of technology in medicine.

Chapters 6–11 are presented in couples, with foundations for the topic being covered in the even chapters and more advanced strategic aspects being covered in the odd chapters: the Food and Drug Administration in Chapters 6 and 7, clinical research in Chapters 8 and 9, and reimbursement in Chapters 10 and 11. The material is prepared so that you may skip the odd-numbered chapters if you feel that the material is too advanced for your students.

Chapters 12–15 address current and practical design challenges. They will help to deepen your students' appreciation for nuances in the design of a device that can easily be overlooked and can affect the safe and efficacious outcome of a medical device.

Chapter 12 develops foundations for infection-prevention design strategies. This chapter was written two years prior to the COVID-19 pandemic. Still, even prior to this event, healthcare-acquired infections and infections due to medical devices were so prevalent that any student of medical devices should understand these issues. The lessons in Chapters 13–15 address considerations for biocompatibility, manufacturing, and use cases of medical devices. These chapters also illustrate challenges in designing medical devices for long-term postmarket safety.

Chapter 16 looks at emerging challenges in medical device design. It reviews many of the issues and lessons that came to light in the text as they apply to current and future medical device design needs, challenges, and opportunities. A discussion of the challenges related to medical devices and technology during the onset of the COVID-19 pandemic is presented to exemplify the summary.

CHAPTER 1

Just 100 Years Ago

Learning Objectives

Understand the crude state of medicine and medical technology just 100 years ago.
Appreciate the significance of developments that have led to our current standard of care.
Recognize the need for accreditation, regulation, and public health policies.
Establish a foundation for the rationale of studying key events that have led to current practices.
Summarize the progress in healthcare and healthcare technology over the past 100 years.
Describe the greatest needs in healthcare today based on (a) costs and (b) causes of death.
Provide an introduction to the business and practice of medical device design.

New Terms

Journal of the American Medical Association (*JAMA*)
new molecular entities (NMEs)
gross domestic product (GDP)

If You Were Born Just 100 Years Sooner

If you were born just 100 years sooner, you most likely would have been born at home—not in a hospital. Your expecting mother would not have told anyone that she was pregnant because it was not an appropriate subject to discuss publicly. If you survived infancy, you would have been lucky because one of every ten infants died by one year of age. If you lived past 47 years, you would have exceeded the average life expectancy.

The four leading causes of death in the United States were, in order, pneumonia/influenza, tuberculosis, diarrhea, and heart disease. These diseases accounted for nearly 40 percent of all deaths (Centers for Disease Control and Prevention [CDC] 1997). There were no antibiotics or vaccines for pneumonia, flu, or infections. There was not a vaccine for tuberculosis; not much was known about it. There were no tests or treatments for digestive diseases or cancer. Diseases of the heart were very poorly understood.

Waterborne disease outbreaks were commonplace. Approximately 25,000 people died each year from typhoid or typhoid-like fevers. Water was not chlorinated; that process was introduced in 1908 and not practiced in a widespread manner. Fewer than 2 percent of people in towns and cities received filtered water, and there was little to no municipal sewerage (Hinman 1990). Any practices of public health were in their infancy.

Hospitals and physicians were not regulated. When the American College of Surgeons was founded in 1913, it was the first organization to offer accreditation to hospitals. During its first review, fewer than 15 percent of hospitals earned accreditation (Moseley 2008). In that period, only half of the medical schools in the United States were affiliated with colleges and universities. At the time, a *Journal of the American Medical Association* (*JAMA*) editorial commented strongly regarding the qualifications of a medical school entrant: "[W]e hope the time will soon come when no person will be permitted to enter upon the study of medicine without presenting proof of a good literary and scientific education" (Hinman 1990).

Physicians traveled to patients' homes, but there was little they could do for most illnesses. There were virtually no medical devices. Technology was primitive. Blood pressure measurement and x-ray technology were just being discovered. Surgical intervention did not even begin to take shape until 1910 (Moseley 2008). Surgical instruments were crude and often designed for other purposes and then adapted for surgical use. Antiseptic principles had only recently been introduced.

How Things Have Changed

U.S. healthcare has progressed tremendously in the past century. A hundred years ago, 40 percent of all deaths occurred in children under five years of age. Today that number is only about 6 percent, and infant mortality has dropped from 10 to 0.6 percent (Thompson 2016). Only a fraction of a percent of childbirths takes place outside a hospital. The average life expectancy in the United States has increased 30+ years to nearly 78 years of age! It is estimated that there are approximately 1,500 different **new molecular entities (NMEs)**, which are drugs that have been approved by the Food and Drug Administration (FDA). The size of the U.S. medical device market has grown from virtually nothing to approximately $156 billion.

Despite all the advances in medicine and the explosive growth of medical technology, we still face many similar health challenges. The following 15 leading causes of death accounted for 80 percent of all deaths in the United States in 2017:

1. Diseases of heart (heart disease)
2. Malignant neoplasms (cancer)
3. Accidents (unintentional injuries)
4. Chronic lower respiratory diseases
5. Cerebrovascular diseases (stroke)
6. Alzheimer's disease
7. Diabetes mellitus (diabetes)
8. Influenza and pneumonia
9. Nephritis, nephrotic syndrome, and nephrosis (kidney disease)
10. Intentional self-harm (suicide)
11. Chronic liver disease and cirrhosis
12. Septicemia
13. Essential hypertension and hypertensive renal disease (hypertension)
14. Parkinson's disease
15. Pneumonitis due to solids and liquids (Heron 2019)

Note that we face many challenges today that are similar to those faced 100 years ago. For example, while the incidence of typhoid fever has been low since the 1940s, other types of infectious diseases contribute to sepsis remaining in the top 20; it ranks twelfth on today's list.

A comparison of the top 10 causes of death from 1900 to 2010 is presented in Figure 1.1. Heart disease and pneumonia/flu remain leading causes of death; heart disease has gone from number four to the number one cause of death.

Our nation has invested significantly in the advancement of healthcare. The good news is that we offer some of the best healthcare in the world, and everyone is entitled to good healthcare. The downside is the cost. We spend almost 20 percent of our **gross domestic**

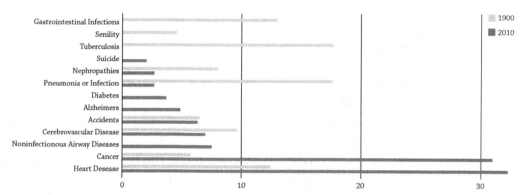

FIGURE 1.1   Top 10 causes of death in 1900 and in 2010.
Values are in percentages of the top 10 causes, where all 10 causes added together equal 100 percent.
Source: Jones, Podolsky, and Greene 2012.

product (GDP) on healthcare, which was nearly $3.5 trillion in 2017 (Centers for Medicare and Medicaid Services [CMS] 2015).

Nearly half of all hospital costs in 2013 were due to these 20 conditions (Table 1.1). If we were able to eliminate just these 20 conditions, we could save the United States approximately $182 billion a year. Imagine what other problems we could solve with an extra $182 billion a year!

There is one thing we have learned about spending and healthcare: *more spending is not necessarily better.* In fact, the United States spends 50 percent more on healthcare than any other developed country. But, compared with most of those same developed countries, life expectancies are lower and infant mortality rates are higher in the United States (Moseley 2008). Better and lower-cost medical devices certainly would contribute to more cost-effective outcomes.

Important Design Knowledge

Throughout the upcoming chapters, we will look closer at many of the conditions outlined in Table 1.1. Because these diseases are so prevalent and so expensive, there is huge value in any good solution. There would also be a tremendous business opportunity for that solution. This is why many companies are focused on these disease conditions in attempts to develop their technologies to solve any of them. If you are working in the field of medical devices, it is very likely that you are working to address at least one of these conditions.

TABLE 1.1   The 20 Most Expensive Conditions Treated in U.S. Hospitals, All Payers, 2013

Rank	CCS Principal Diagnostic Category	Aggregate Hospital Cost in Billions ($)
1	Septicemia	23.66
2	Osteoarthritis	16.52
3	Liveborn	13.28
4	Complicaton of device, implant or graft	12.43
5	Acute myocardial infection	12.09
6	Congestive heart failure	10.21
7	Spondylosis, invertebrate disc disorders, other back problems	10.19
8	Pneumonia	9.50
9	Coronary atherosclerosis	9.00
10	Acute cerebrovascular disease	8.84
11	Cardiac dysrhythmias	7.17
12	Respiratory failure	7.07
13	Complications of surgical procedure	6.07
14	Rehabilitative care, fitting of prosthesis, and adjustment of device	5.37
15	Mood disorders	5.24
16	Chronic obstructive pulmonary disease	5.18
17	Heart valve disorders	5.15
18	Diabetes mellitus with complications	5.14
19	Fracture of neck or femur (hip)	4.86
20	Biliary tract disease	4.72
	Total	**181.69**

Note: These 20 conditions accounted for half of all hospital costs. A more cost-effective treatment for any of these conditions would make a huge financial impact on our hospital systems.

Source: Torio and Moore 2016.

We will review the nature of these conditions, current solutions, and challenges to improving solutions for each. You will learn that some of those challenges are in technology, but many are related to other issues, such as regulatory concerns, quality, equipment management, and user acceptance, just to name a few. To be successful, a biomedical engineer must know how to introduce technology that will satisfy both the technical and nontechnical requirements.

Summary

Clearly, much has changed in the past 100 years. Technology has exploded, and medicine has become very sophisticated. American lifestyles have grown to be more complicated, expensive, and regulated. The good news is that we have been able to take advantage of these advancements to provide amazing technological solutions for healthcare.

But not every step forward was achieved without a price. As we advanced technology to improve medicine and healthcare, we learned invaluable lessons—sometimes at the expense of causing illness, injury, or even death to those for whom we strived to provide care. The lessons learned were costly but have greatly contributed to the evolution of today's practices both in the delivery of medical care and in the application of technology to support it.

If only there were a study guide to all that we have learned during the development of medical technology over the past 100 years, it would be so much easier to understand and apply the seemingly anomalous rules in medical device/technology design. Luckily for you, this book is that study guide. It will help you to learn about critical events that shaped our current practices. It will help you to appreciate current healthcare challenges. The lessons and examples in this book will help you to navigate both the technical and nontechnical issues that you must consider in medical device design. Study this material well, and you will gain the background you need to make good, confident decisions in your daily design practices.

Study Questions

1. What is tuberculosis? When was the vaccine for it discovered?
2. How did typhoid fever spread?
3. What were the criteria for entrance into medical school prior to 1900?
4. For what reason(s) was it that until the 1900s the likelihood of surviving surgery was greater than the chance of dying during or soon after?
5. How large is the medical device market in the United States?
6. Which diseases or conditions have been leading causes of death for 100 years?
7. Which diseases or conditions were not leading causes of death 100 years ago?
8. Why is it important to review the most expensive conditions treated in a hospital?
9. What is one way that you might reduce the 20 percent of U.S. GDP spent on healthcare?

Thought Questions

1. Who was Joseph Lister, and why was he important? Could you call him a biomedical engineer? Why or why not?
2. What led to the development of the first antibiotics?
3. When was the first successful heart surgery performed? Where did it take place, and what was the nature of the procedure?
4. How have the leading causes of death changed compared with today? In what way are they similar (see Figure 1.2)?

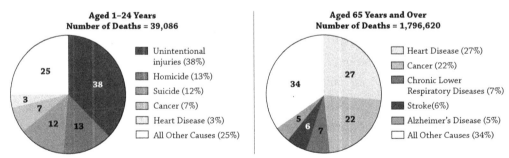

FIGURE 1.2   Top five causes of death in the United States, separated by age.
Source: Kochanek et al. 2019.

5. What differences do you notice in Figure 1.2 comparing causes of death in young versus old population? How would you imagine this figure has changed since the 1900s? How do you think the differences in this figure contribute to healthcare costs?

6. Look up some charts on average life expectancy in the United States. Limit your search with specifics such as gender, race, region of the United States, or socioeconomic status. What do you notice? What could biomedical engineers do to change this?

References

Centers for Disease Control and Prevention (CDC). 1997. "Achievements in Public Health: Control of Infectious Diseases." *Morbidity and Mortality Weekly Report* 48.

Centers for Medicare and Medicaid Services (CMS). 2015. "NHE Fact Sheet, 2013," 2014–2015. Available at www.cms.gov/research-statistics-data-and-systems/statistics-trends -and-reports/nationalhealthexpenddata/nhe-fact-sheet.html.

Heron, Melonie. 2019. "Deaths: Leading Causes for 2017." *National Vital Statistics Reports* 68(6):1–76.

Hinman, A. R. 1990. "1889 to 1989: A Century of Health and Disease." *Public Health Reports* 105(4):374–380.

Jones, David S., Scott H. Podolsky, and Jeremy A. Greene. 2012. "The Burden of Disease and the Changing Task of Medicine." *New England Journal of Medicine* 366(25):2333–2338. https://doi.org/10.1056/NEJMp1113569.

Kochanek, Kenneth D., Sherry L. Murphy, Jiaquan Xu, and Elizabeth Arias. 2019. "Deaths: Final Data for 2017." *National Vital Statistics Reports* 68(9). www.cdc.gov/nchs/products/ index.htm.

Moseley, George B. 2008. "The U.S. Health Care Non-System, 1908–2008." *Virtual Mentor* 10(5):324–331. https://doi.org/10.1001/virtualmentor.2008.10.5.mhst1-0805.

Thompson, Derek. 2016. "America in 1915: Long Hours, Crowded Houses, Death by Trolley." *The Atlantic*, February 2016. www.theatlantic.com/business/archive/2016/02/ america-in-1915/462360/.

Torio, Celeste M., and Brian J. Moore. 2016. "National Inpatient Hospital Costs: The Most Expensive Conditions by Payer, 2013," Statistical Brief No. 204. In *Healthcare Cost and Utilization Project (HCUP) Statistical Briefs*. https://doi.org/10.1377/hlthaff.2015.1194.3.

CHAPTER 2

President Eisenhower's Heart Attack

Learning Objectives

Review the development of biomedical technology in the last century via two important examples: electrocardiograms (ECGs) and oxygen (O_2) therapy.

Provide the historical context of President Eisenhower's heart attack in the landscape of standards of care for a heart attack.

Understand the basic concepts of myocardial infarction (MI).

Answer questions around the beginnings of biomedical engineering as we know it today:

Who were some notable forefathers of biomedical engineering?

What are the types of disciplines involved in biomedical engineering?

How are ideas shared/applied to medicine via interdisciplinary work?

Understand the state of drug treatment strategies associated with medical device applications.

Understand the current costs of MI.

New Terms

amyl nitrite snuff

papaverine

morphine

electrocardiogram (ECG) machine

myocardial infarction (MI)

heparin

Einthoven's triangle

translation

efficacious

hyperbaric oxygen

September 1955

In September 1955, while playing golf, President Eisenhower began to complain about an upset stomach. He attributed it to indigestion that was likely caused by his lunch, a hamburger with Bermuda onion slices, and went home for the day not thinking much of it. Just after midnight, however, he awoke with severe chest pain and asked his wife for milk of magnesia. This was a common antacid at the time—it would be like asking for Tums or Pepto Bismol today. Mamie, his wife, was worried and called his personal physician, Dr. Snyder (Lasby 1997).

Dr. Snyder went to their home at about 2 a.m. and treated the president with **amyl nitrite snuff** (this was a common treatment to relieve chest pain). He also gave President Eisenhower injections of **papaverine** (which was used to treat spasms of the gastrointestinal tract) and **morphine** (to relieve pain). Morphine has a tendency to cause sleep. When the President awoke at 11 a.m. with persisting chest pains, arrangements were made to connect him to an **electrocardiogram (ECG) machine**. The White House had to engage a cardiac specialist, Dr. White, to bring the machine from a local hospital to the president and to operate it. The diagnosis, which took nearly 24 hours since the onset of his first symptoms, revealed a **myocardial infarction (MI)**—a heart attack! The president was taken to the hospital. He was placed in an oxygen tent, treated with **heparin** (an anticoagulant) to stabilize his blood flow and ordered extended bed rest (Lasby 1997).

By the middle of the twentieth century, modern medicine was coming of age, and President Eisenhower received the best care possible. He was treated by some of the nation's best doctors with the latest drugs, medical protocols, and state-of-the-art medical technology. A closer look at that event will provide you with some appreciation for the foundations of our practices today, as well as for how physicians, biomedical engineers, and scientists of the twentieth century have contributed to advancing the standard of care in this field.

Electrocardiogram (ECG)

You might be thinking that Dr. Snyder's midnight house call to President Eisenhower was special treatment. After all, if you called your doctor at midnight complaining of chest pain, he or she would certainly not come to your house. Believe it or not, it was not unusual for a physician to make house calls at that time. Keep in mind that there was no 9-1-1 service and there were far fewer ambulances. In fact, especially in rural locations, a house call was the most effective—and sometimes the only—way for a patient to see a doctor.

While the physician's house call might not have been special treatment, having an ECG performed at home certainly was. President Eisenhower might have been the first patient to ever have had an ECG machine brought to his home. ECG machines in that day were not nearly as commonplace as they are today. While somewhat portable, they were big and bulky, weighing nearly 30 pounds, and they required expertise to operate (Figure 2.1). Still, ECG machines had developed rapidly since the early 1900s, when they weighed 600 pounds and took five people to operate (Figure 2.2).

FIGURE 2.1   Portable ECG machine resembling the one brought to President Eisenhower's bedside in September 1955.

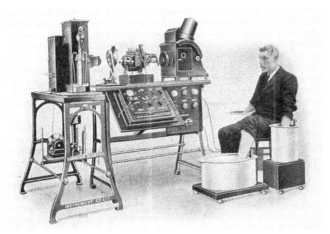

FIGURE 2.2   String galvanometer ECG machine circa 1901.

Physician/physicist Willem Einthoven paved the way for the practical use of the ECG machine in the clinic. Inspired by the earlier research of physiologist August Waller, Einthoven improved on Waller's five-lead capillary electrometer, which he used to study phases of electrical activity associated with cardiac rhythms. Einthoven built a string galvanometer that improved on the accuracy of the capillary electrometer. He was able to reduce the need for five leads down to three, removing one in the mouth and another on the right leg from Waller's design (AlGhatrif and Lindsay 2012). He created electrodes by immersing the subject's hands and foot in large buckets of saline, as shown in Figure 2.2.

Einthoven, who was very comfortable with electrical physics as well as medicine, conceptualized his three-lead system as an inverted equilateral triangle superimposed over the body with the heart at the center and one lead at each corner of the triangle, as shown in Figure 2.3. A mathematically elegant approach, his concept, **Einthoven's triangle**, enabled him to predict the direction of the cardiac muscle cell depolarization using measured deflections from any two leads, independent of the position of the heart's electrical vector. With this "simplification," cardiac rhythms could be studied for clinical value—to determine whether a patient's rhythms were normal or not. He was awarded the Nobel Prize in 1924 for the invention of the ECG machine (AlGhatrif and Lindsay 2012).

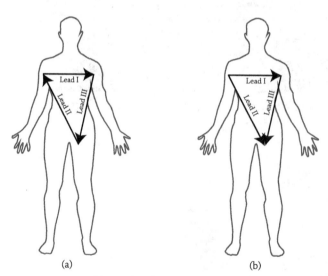

(a)   (b)

FIGURE 2.3   (a) Kirchoff's loop rule states that the sum of all potential differences around a closed loop is zero. (b) Einthoven reversed lead II such that lead I + lead III = lead II. Thus, by measuring the values for any two of the leads, the value for the third lead can be calculated.

The first clinical use of the ECG machine in the United States was at Mt. Sinai Hospital, New York, in 1909. By 1930, the ECG had become so well studied that it could be used to differentiate cardiac from noncardiac chest pain. In some cases, patterns were so distinctive that they could be used alone to confirm the diagnosis of MI (AlGhatrif and Lindsay 2012). In just 30 years, the ECG machine had evolved from a heavy, cumbersome piece of lab equipment used for experimentation to a portable device that is useful in many clinical diagnoses. This was a remarkably swift **translation**, even by today's standards.

Oxygen Tent

The oxygen tent that President Eisenhower was placed in when he arrived at the hospital is similar to the one shown in Figure 2.4. This tent to help the president breathe more easily and take the strain off his heart was quite an advanced system from the first ones that were introduced to practice around the 1920s. Because plastic was not common at that time, the original tents had opaque, rubberized curtains. The oxygen was delivered along with a tremendous amount of heat, making it unbearably hot within the tent. The discomfort was so intense that some practitioners considered the treatment inhumane, even calling for a ban on it. But early innovative caregivers brought ice with them to accommodate patients, and ultimately, the tents were engineered with refrigeration systems included. The development of plastics allowed for other innovations, such as clear, lightweight curtains so that patients could see what was happening outside the tent. President Eisenhower certainly benefited from early biomedical engineering.

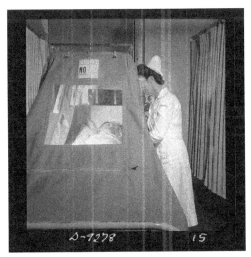

FIGURE 2.4   An oxygen tent and the accompanying stand-alone oxygen tank circa 1942. Engineers later built an oxygen storage and refrigeration cabinet to reduce the heat generated by the system. Oxygen therapy also posed a risk of fire. The sign on the clear plastic curtain says, "No Smoking."

Oxygen Therapy

Oxygen tents are antiquated by today's standards. The new standard of care is oxygen therapy, used so pervasively in our contemporary medical practices that we almost take it for granted. But it took centuries of interdisciplinary scientific and medical research and engineering to bring oxygen therapy to the **efficacious** levels that we know today.

The first reported use of therapeutic oxygen was in 1783, when a young woman with tuberculosis was treated with daily inhalations and was said to have her condition greatly improve. This prescribed regimen was fewer than 10 years after oxygen was discovered independently by pharmacist Karl Scheele and chemist Joseph Priestley (Grainge 2004). But it would take another 100 years before oxygen therapy would become a commonly accepted practice in medicine.

In the 1800s, oxygen therapy was touted by some and dispelled by others. The medical community did not have a good understanding of how the body used oxygen, and discussions in publications revolved around its merit, dosage quantities, and delivery methods. Some practitioners expressed concerns about excessive dosages: "[N]o human being could possibly stand so great an amount of oxygen on account of the dangerous degree of stimulation to the system and the increased combustion of tissue" (Blodgett 1890). Others proposed that oxygen be administered "via stomach for resuscitation, urethra for inflammatory conditions and enema for treatment of gallstones" (Grainge 2004). Most therapies were delivered intermittently and, by today's standards, used very small amounts of oxygen, offering little efficacy.

Some meaningful breakthroughs emerged near the turn of the twentieth century. In 1888, Dr. Albert Blodgett reported in the *Boston Medical and Surgical Journal* a new way to deliver oxygen therapy. He described a "continuous" delivery of oxygen versus the traditional couple of gallons at a time and reported great success in his pneumonia patient (Blodgett 1890). Around the beginning of the 1900s, physiologists Adolph Fick and Paul Bert identified the concept of partial pressure. They used it to describe the difference in oxygenation between arterial and venous blood. This laid the foundation for rapid adoption and advancement of oxygen therapy.

Building on this foundation, by 1917, Dr. John Scott Haldane had published a paper on therapeutic oxygen administration. He provided a detailed explanation of how respiration is driven by carbon dioxide and how it affects blood hydrogen ion concentration. His descriptions were so accurate that not much has changed in our understanding even today (Grainge 2004).

We learned a lot more about oxygen therapy during World War I because the poisonous gas phosgene was used as a chemical weapon, and oxygen was used to treat soldiers who fell victim to it. Young medical officers working in the battlefields of France employed Haldane equipment, a pressurized cylinder, pressure regulator, and a tight-fitting mask, to deliver oxygen treatment, as shown in Figure 2.5.

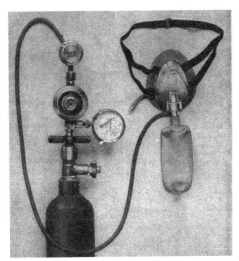 

FIGURE 2.5   Haldane's early oxygen therapy apparatus.

Wartime necessity led to examples of biomedical engineering in the battlefield. Lung edema associated with phosgene poisoning caused the victims to expurgate to fill the mask, rendering the mask ineffective. On seeing this, Captain Adrian Stokes, a field medical officer, developed nasal prongs—the forefather of the nasal cannula that is used today (Grainge 2004).

A Half Century after the President's Heart Attack

In fewer than 50 years after President Eisenhower's heart attack, emergency care for everyone in the United States has been built into the core of our community's infrastructure. Medical equipment has evolved rapidly, embracing all that Silicon Valley and the telecommunications and biotechnology industries have to offer.

Today, almost everyone knows that speedy treatment is critical to surviving an MI, and even the layperson knows that any complaint of chest pains should not be overlooked. You or someone nearby would call 9-1-1, and within minutes, an ambulance would be at your door. Paramedics would bring a portable ECG machine that weighs only a pound or two into your home. They would quickly place some disposable sticky electrode leads on you and could even transmit your heart rhythm information to the hospital in almost real time, where surgeons would be standing by, preparing an appropriate intervention for you. You would be treated with oxygen immediately from a small portable oxygen supply with a comfortable and effective nose cannula or mask.

In this day and age, such special treatment is a basic standard of care. To think that the president of the United States had a doctor who originally missed the diagnosis is inconceivable.

You can even buy a personal ECG machine—a small, inexpensive ECG device that weigh less than a pound, has no wires, and synchs to your smart phone. These devices allow for patients to run their own ECGs, storing the data for personal use or transmission to medical professionals. Although this small personal device may not be as sophisticated as some larger professional devices, it certainly offers more information in just a matter of seconds than the device used on President Eisenhower by a field expert at the time.

You will not find oxygen tents in use today at all for adults, only perhaps in a few conditions involving small children. Small, comfortable oxygen systems are readily available for emergency care, as well as for safe in-home long-term therapy. Even oxygen generators are now portable.

Foundations for Medical Device Advancement

While standard of care for MI has certainly progressed since 1955, it is interesting to note just how much we already knew by then. In fact, we still use many of the same drugs and therapies as President Eisenhower received. Certainly, we know more about the mechanisms of action for many drugs, we have developed a better understanding of dosages, and equipment is much more advanced. But the foundations for care were well established by the many brilliant scientists and clinicians of previous centuries who were dedicated to improving patient care and outcomes.

With this foundation laid, and with the postwar U.S. technology enterprise, doors opened widely for technological innovation in medicine. Physicians, scientists, and engineers collaborated in teams to solve medical problems with the latest technology. Likewise, they joined to advance technology to solve a pressing medical need. Indeed, biomedical engineering was flourishing many years before it became recognized as its own field of study in the early 1960s.

Ike: Resilience Thanks to Biomedical Engineering

In the 1950s, a diagnosis of heart attack was considered a death sentence. But President Eisenhower and his doctors ushered in a new era for heart patients. The president recovered fully from that first heart attack and went on to serve a second term. He had seven more heart attacks and, in each case, received the most up-to-the-minute treatment available. He continued to benefit from the introduction of other new drugs and equipment, including the early direct-current (DC) defibrillator.

There are varying accounts of the first 24 hours of President Eisenhower's heart attack, and we will never know the entire truth for certain. Some people suggest that Dr. Snyder knew that the president was having a heart attack but chose not to disclose it to the president's family or the public so as not to cause alarm. Other accounts describe how plausible it was to confuse the president's symptoms with indigestion. The ECG results would have been the only way to tell for certain. In any case, this event and the discussions about heart disease that surrounded it were pivotal in our development of methods and technology to improve the outcomes of heart attack patients.

Summary

You might feel as if we have seemingly mastered the care for heart attacks and be wondering what else there is do. How could you possibly contribute to advancing care for MI? Unfortunately, despite all the advances we have made, MI is still one of the top five most expensive conditions during inpatient hospitalizations in the United States, with a cost of about $12.1 billion, accounting for over 600,000 hospital stays annually (Torio and Moore 2016) .We still have a long way to go to help prevent heart attacks, treat them more effectively, and reduce the costs of care. New biomedical engineering will be essential in these efforts.

One important final note about President Eisenhower's case: after his first heart attack in 1955, his doctors summarized what they believed were contributing risk factors. They grouped the factors into what they believed at the time to be the most important risk factors for MI. These included older age; male sex; broad, muscular body build; active, ambitious personality; and heredity. Further, they listed what they believed to be environmental contributors, including stress, diet, and exercise. The doctors included notes about factors that needed appraisal but were considered to be "much less important," and those included alcohol use (Messerli, Messerli, and Lüscher 2005).

Today we know a lot more about the risk factors for heart disease, thanks to the half century of study that followed these announcements. President Eisenhower was an avid golfer, which led some to believe that exercise and heart disease were linked, and we know now this is incorrect. The president was also known to smoke two packs of cigarettes a day and to consume a diet that was "very unhealthy" by today's standards. So his doctors were on the right track with these assessments. We also know that alcohol use is a very important risk factor. This demonstrates how medical ideas and practices can change as we study diseases. It is also a reminder to all of us about how we can prevent heart disease: *eat right, exercise, and don't abuse alcohol, smoke, or vape!*

While oxygen therapy is already used in so many ways, researchers continue to study how oxygen can be applied to heal the body, such as using **hyperbaric oxygen** for the treatment of wounds and other conditions. Much like oxygen treatments in the 1800s, it is still unclear how to select and deliver effective dosages of hyperbaric treatments for some conditions, but the potential is there for some major breakthroughs from research and biomedical engineering. Interestingly, the laws of partial pressure, introduced by Fick and Bert, can be used to describe how hyperbaric chamber treatments provide delivery of three times as much oxygen to the body compared with atmospheric pressure.

Biomedical engineering is firmly entrenched in the cycle of advancing the human condition. Just as the forefathers Einthoven and Haldane made broad contributions and Stokes made a specific improvement, biomedical engineers are an integral part of the team dedicated to saving lives, reducing medical costs, and expanding access to better healthcare. We will always have more work to do. It's a big job, but you've got this!

Study Questions

1. In what field was Willem Einthoven's formal training? Would you consider him to be a biomedical engineer? Why?
2. How was Einthoven able to predict the direction of the cardiac muscle cell depolarization?
3. What did Dr. Blodgett do differently to make oxygen therapy successful?
4. How did Fick and Bert describe the difference in oxygenation between arterial and venous blood?
5. What key ideas about oxygen therapy did Dr. Haldane introduce?

Thought Questions

1. List five things that are different today compared with 1955 for the diagnosis and treatment of heart attack. List five things that are similar. Is there something you think that even with all the technology we have today we must do better?
2. Draw a timeline describing the size/weight and key features of cardiac rhythm monitoring equipment from 1900 to today.
3. Describe some of the major changes in oxygen therapy from the late 1800s to today.
4. Look up images of oxygen tents from 1900, 1930, 1960, 1990, and today. How have they changed?

References

AlGhatrif, Majd, and Joseph Lindsay. 2012. "A Brief Review: History to Understand Fundamentals of Electrocardiography." *Journal of Community Hospital Internal Medicine Perspectives* 2(1):14383. https://doi.org/10.3402/jchimp.v2i1.14383.

Blodgett, Albert. 1890. "The Continuous Inhalation of Oxygen in Cases of Pneumonia Otherwise Fatal and in Other Diseases." *Boston Medical and Surgical Journal* 123:481–485. www.nejm.org/doi/full/10.1056/NEJM189011201232101.

Grainge, C. 2004. "Breath of Life: The Evolution of Oxygen Therapy." *Journal of the Royal Society of Medicine* 97(10):489–493. https://doi.org/10.1258/jrsm.97.10.489.

Lasby, Clarence G. 1997. *Eisenhower's Heart Attack: How Ike Beat Heart Disease and Held on to the Presidency*. Lawrence: University of Kansas Press.

Leigh, Julian M. 1974. "The Evolution of Oxygen Therapy Apparatus." *Anaesthesia* 29(4): 462–485. https://doi.org/10.1111/j.1365-2044.1974.tb00688.x.

Messerli, Franz H., Adrian W. Messerli, and Thomas F. Lüscher. 2005. "Eisenhower's Billion-Dollar Heart Attack: 50 Years Later." *New England Journal of Medicine* 353(12):1205–1207. https://doi.org/10.1056/NEJMp058162.

Torio, Celeste M., and Brian J. Moore. 2016. "National Inpatient Hospital Costs: The Most Expensive Conditions by Payer, 2013," Statistical Brief No. 204. In *Healthcare Cost and Utilization Project (HCUP) Statistical Briefs*. https://doi.org/10.1377/hlthaff.2015.1194.3.

CHAPTER 3

Events That Built Our Key Regulations and Practices

Learning Objectives

Provide a summary of milestones in U.S. history relevant to the foundation of common regulatory and quality-management practices encountered in today's medical device design.

Show the rationale and evolution of U.S. regulations and practices.

Understand current regulations and practices in medical device design.

Deliver an accurate description of the scope and purpose of current regulations and practices.

Establish an appreciation for the policy drivers so as to navigate the regulation or practice more effectively.

New Terms

United States Pharmacopeia (USP)

Food and Drug Administration (FDA)

Food, Drug, and Cosmetic Act (FDCA)

good manufacturing practices (GMPs)

National Institutes of Health (NIH)

current good manufacturing practices (CGMPs)

FDA 21 Code of Federal Regulations (CFR), Part 820: Quality System for Medical Devices

Medical Device Amendments Act of 1976

Manufacturer and User Facility Device Experience (MAUDE) database

Safe Medical Device Act of 1990

design control

premarket approval (PMA)

International Organization for Standardization 13485 (ISO 13485)

Food and Drug Administration and the Food, Drug, and Cosmetic Act of 1938

America has been addressing food safety concerns since the colonial period. As early as 1641, the states of Massachusetts and Virginia required bakers to place their identifying mark on their bread. In 1813, the first federal biologics law was passed; it addressed the provision of reliable smallpox vaccine to citizens (U.S. Food and Drug Administration [FDA] 2006). This was enacted to encourage people to get vaccinations against smallpox. While concerns over food and drug quality were evident early in the United States, it would take nearly another century of evolution before the government would establish an enforceable effort to regulate and control food and drug quality and accessibility.

By the early nineteenth century, the United States had become a dumping ground for counterfeit, diluted, and decomposed drug materials that were unmarketable in Europe. American troops in Mexico were facing high casualties that were attributed to the administration of weak and adulterated drugs for the treatment of malaria. U.S. federal control over the drug supply began in 1848 with the Drug Importation Act, which required U.S. Customs Service inspection to stop the entry of adulterated drugs from overseas. The act required imported drugs to meet the standards for strength and purity established in the *United States Pharmacopeia* (*USP*), established by trade and professional leaders in 1820. Gradually, support faded and so did the program (U.S. FDA 2006).

The formal roots of the **Food and Drug Administration (FDA)** go back to 1862, when President Lincoln appointed a chemist, Charles M. Wetherill, to serve in the new Department of Agriculture. The department established the Bureau of Chemistry, where Dr. Wetherill set up a laboratory to analyze samples of food, soils, fertilizers, and other agricultural substances (U.S. FDA 2006). Dr. Wetherill was succeeded by other chemists, Dr. Peter Collier and then Dr. Harvey Wiley, who worked hard to counter adulteration practices.

Dr. Wiley established his reputation as the "crusading chemist." During the late 1800s and early 1900s, he campaigned for federal law to address food adulteration that he investigated, though he faced strenuous opposition from whiskey distillers and the patent medicine firms, for fear that they would be put out of business by federal regulation. Furthermore, it was argued that the federal government had no business policing what people ate, drank, or used for medicine (U.S. FDA 2006).

The Pure Food Bill

Meanwhile, in the 1870s, the pure food movement began to take shape. Originally, food industry members advocated for a federal law against adulteration. They were concerned about competition from a new breed of food products, that is, glucose as a replacement for sugar, "lard" made from cottonseed oil, and oleomargarine as a substitute for butter. Also, food packers were finding it difficult to manufacture products that met the variations in laws among states.

By 1905, surveys of food examinations indicated that more than half of food samples were below standard. Products sold as olive oil contained cottonseed oil instead. There was alcohol in candy and medicines. Medicines with false or exaggerated claims often contained narcotic or addictive drugs such as cocaine, opium, and morphine, but these ingredients were not included on the label. Some of these medicines were recommended for soothing children but often proved fatal. The *Congressional Record* tallied many reports of the harmful effects of these products (Barkan 1985).

Ultimately, when the activist clubwomen of the country rallied to the pure food cause, political support amassed in its favor. At long last, the Pure Food Bill, "[f]or preventing the manufacture, sale, or transportation of adulterated or misbranded or poisonous or deleterious foods, drugs, medicines, and liquors and for regulating the traffic therein, and for other purposes," reached Congress in 1905 ("The Pure Food Bill" 1906). Boosted by public outcry after exposure of insanitary conditions in the Chicago meat packing industry in Upton Sinclair's 1906 book, *The Jungle*, both the Pure Food Bill and the Federal Meat Inspection Act were enacted into law on the same day.

Finally, in 1906, President Theodore Roosevelt signed into law the first major food and drug safety bill, the Pure Food and Drug Act. (Just in case you needed a refresher from seventh grade social studies, recall that when a bill is passed by Congress and signed by the president, it becomes a law. An *act* is a single enacted bill and refers to one type of law. A law may be broader or incorporate multiple acts.) The Pure Food and Drug Act was amended in 1938 to the **Food, Drug, and Cosmetic Act (FDCA)** that we know today. In 1927, the Bureau of Chemistry became the United States Food, Drug, and Insecticide Administration; the name was shortened in 1930 to the U.S. Food and Drug Administration (FDA).

Early Good Manufacturing Practices and the National Institutes of Health

Foreshadowing the Pure Food and Drug Act, the Biologics Control Act was passed by Congress in 1902. This act was in reaction to a 1901 tragedy in which 13 children in St. Louis were treated with an antiserum for diphtheria only to die of tetanus. It was discovered that the horse serum used to make the vaccine had been contaminated with tetanus (U.S. FDA 2006).

This legislation gave rise to regulation of the production of vaccines and antitoxins. Between 1903 and 1907, standards were established and licenses were issued to pharmaceutical firms for making smallpox and rabies vaccines, diphtheria and tetanus antitoxins, and various other similar drugs, laying the foundation for guidelines that are known today as **good manufacturing practices (GMPs)**.

The regulatory responsibility was handed to the then recently formed Hygienic Laboratory. This laboratory was initially focused on isolating the organism that caused cholera. Its responsibilities grew to perform selected research for the study and control of infectious diseases. Related duties then included training and support of government health scientists in public health and clinical applications.

After 10 years of work at the Hygienic Laboratory, Director Dr. Joseph Kinyoun proposed that the federal government should create a laboratory-supported national and international research enterprise to "[look] into the nature, origin, and prevention of contagious epidemics, and other diseases affecting the people, and should also make investigations into other matters relating to public health." By 1930, this laboratory became the **National Institutes of Health (NIH)**. Today, the NIH is the largest biomedical research agency in the world and resides within the structure of the U.S. Department of Health and Human Services. Among its many roles, it serves to establish standards for research and healthcare quality. Dr. Kinyoun is fondly referred to as the "Father of the NIH" (National Institutes of Health 2018).

Thus, by the early 1900s, the stage was set for the intricate regulatory and quality-management practices that we apply today for medical device design and manufacturing. These practices have evolved as a result of a series of events and experiences that often had terrible consequences. Best practices often reflect on a precedent event, and you will develop an appreciation for such practices by studying the precedents. The FDA website (www.fda.gov) is a tremendous source for this information.

Elixir of Sulfanilamide

Sulfanilamide was a drug used to treat streptococcal infections. By 1937, it had already been used safely for a while in tablet and powder forms. To meet a new demand for a liquid form of the drug, the company's chief chemist and pharmacist, Harold Cole Watkins, experimented to find what he thought was a good solvent for sulfanilamide: diethylene glycol. The mixture was tested for flavor, appearance, and fragrance, but the new formulation had not been tested for toxicity. At that time, pharmacologic studies were not required (Ballentine 1981).

Diethylene glycol is a chemical that is normally used as an antifreeze. It is also deadly. You may have heard about how this chemical had been a danger to pets and wildlife because of its seductively sweet taste. Animals would drink it and then would become very sick or die. In fact, today, antifreeze manufacturers add a bitter agent to counter its sweet taste.

Watkins may have cleverly selected diethylene glycol as a solvent because of its sweetness, which may have been a benefit to the flavor of an elixir. Apparently, however, he was not aware of its toxicity. His elixir was ultimately responsible for the deaths of more than 100 people in the United States. The deaths led to the passage of the 1938 Food, Drug, and Cosmetic Act, which increased the FDA's authority to regulate drugs. This is only one of many changes to the FDA's authority that were based on tragic events.

Important Design Knowledge

While the FDA has certainly developed many guidelines since 1906 as a result of (often catastrophic) events that occurred over the years, a medical device designer will find that there may be many different ways to design and manufacture a device. Although a device must meet the regulations, the manner in which those regulations are met is entirely up to the designer. Generally, however, the more that a design, method, or process mirrors those which are safely in practice, the less new evidence is needed to support that device's application. Still, there is room for significant innovation and improvement under the FDA's guidelines because they leave the specific manner in which to comply with regulations open.

Later chapters will devote more time to the FDA. Here I offer a brief summary. After proper registration, the FDA implements a review process for a proposed design, and based on its collective experience and knowledge, it determines whether the device may be marketed in the United States. For a new technology for which there is no prior experience with similar devices, the burden is on the device manufacturer to prepare and present supporting evidence and data, usually from clinical trials and rigorous laboratory tests.

The FDA is ultimately responsible for defining an acceptable level of risk and may determine that specific tests must be conducted or that a test must be performed in a certain manner in order to satisfy that concern for risk. It is very important for device designers and manufacturers to work closely with the FDA to ensure that the tests they plan will satisfy the FDA. Ultimately, though, the FDA reviews the entire body of evidence when it is submitted and may make further recommendations and suggestions beyond the original plan.

This collaborative aspect is often the most challenging to a device designer because the regulatory requirements are not specific and require interpretation. Sometimes that interpretation will differ between the manufacturer and the agency evaluator(s), and until that gap is reconciled, product design phases or launch may be delayed. Today, even the most experienced device designers may find it difficult to know exactly what must be done to meet FDA requirements.

Similarly, there are no universal GMPs. Because the regulation must apply to so many different types of devices, the regulation does not prescribe in detail how a manufacturer must produce a specific device. Rather, the regulation provides the framework that all manufacturers must follow by requiring that manufacturers develop and follow procedures and fill in the details that are appropriate to a given device according to the current state-of-the-art manufacturing for that specific device (U.S. FDA, Center for Devices and Radiologic Health 1997).

In fact, it is very possible that two companies that make almost the same product will make them with quite different processes. For example, one company might use gamma radiation for sterilization, whereas the other company could use ethylene oxide. While these are two very different sterilization methods, if both render a sterile product—without damaging the product in that process—either method may be an acceptable GMP.

Current Good Manufacturing Practices and FDA 21 CFR Part 820

Today, GMPs, also referred to as **current good manufacturing practices (CGMPs)**, describe a set of minimum requirements for the design, monitoring, and controls over the manufacturing processes and facilities for a product to ensure that it meets quality standards. The "C" is included to remind us that *current*, up-to-date practices should be adopted, and it is expected that many practices will change and improve over time.

To satisfy GMP requirements, procedures must be documented in writing by the device manufacturer. This contributes to a quality-management system that ensures that the medical device is manufactured to be safe and effective because it helps to prevent instances of contamination, mix-ups, deviations, failures, and errors. Controls must also be placed to detect and investigate product quality deviations, and reliable testing laboratories must be maintained (Lee et al. 2015).

The FDA promulgates GMPs in the United States. GMPs are one of the FDA's set of guidelines that regulate how drugs can be made. The Medical Device Amendments of 1976 prescribed CGMP requirements for the methods used in and the facilities and controls used for the manufacture, packing, labeling, storage, installation, and servicing of all finished medical devices intended for human use. These regulations were enacted on December 18, 1978 (U.S. FDA, Center for Devices and Radiologic Health 1997) and were meant to ensure that medical devices achieve their intended efficacy. Medical device manufacturers undergo FDA inspections to ensure compliance with the current set of guidelines. These can be found in **FDA 21 Code of Federal Regulations (CFR), Part 820: Quality System for Medical Devices** (U.S. Government Publishing Office 2019).

Each manufacturer is responsible to establish requirements for each type or family of devices that will result in devices that are safe and effective. Similarly, the medical device manufacturer must establish methods and procedures to design, produce, and distribute devices that meet the quality system requirements. It is important to note that the responsibility for meeting these requirements and for having objective evidence of meeting these requirements remains with the medical device manufacturer and may not be delegated, despite whether the actual work might be outsourced or otherwise delegated.

Medical Device Amendments Act of 1976

To start, it is useful to know how a medical device is described today. Briefly, a *medical device* is a product, such as an instrument, machine, implant, or in vitro reagent, that is intended for use in the diagnosis, prevention, or treatment of diseases or other medical conditions. A specific definition of a medical device may be found on the FDA website (www.fda.gov).

In 1938, medical devices were relatively simple. By the 1960s, however, they had become far more complex. Basic instruments of the early 1900s had evolved into high-tech products such as heart valves, x-ray machines, and others. Although the FDC extended federal control to devices in 1938, the regulations in that document were ill-suited for current medical devices, and there was a growing concern for better oversight. The FDA had no regulatory power to keep a device off the market; instead, it was required to initiate action against a product *after* it was determined to be misbranded or adulterated (Hyman 1992).

In the late 1960s, with staunch urging from President Richard Nixon, the Cooper Committee was commissioned to study the adverse effects of medical devices for human use (see Chapter 5). In 1970, the committee recommended a classification for medical devices based on relative risk. It was recommended that regulations be carefully tailored to the type of risk a device could impose. The recommendations were adopted in the **Medical Device Amendments Act of 1976**. With these amendments, the FDA was granted authority to effectively control medical devices. The amendments included a requirement that the FDA be notified of every medical device prior to its marketing, and device establishments were required to register with the FDA. Recommendations authorizing GMPs were also included.

Safe Medical Devices Act of 1990 and Design Controls

The **Safe Medical Devices Act of 1990** grew out of congressional concerns about the FDA's ability to quickly learn when a medical device caused an adverse patient event and to ensure that hazardous devices were removed from hospitals and other healthcare facilities in a timely manner. Users had always had the option to voluntarily report adverse events, but few reports were filed until mandated in 1991. This act was established to give the FDA the legal authority to directly regulate the use of medical devices in healthcare facilities. It includes specific requirements for hospitals, health professionals, and other users of medical devices to report patient incidents involving medical devices to the manufacturer and to the FDA if a device caused or contributed to a serious injury, death, or other adverse experiences. *Adverse experiences* are defined by the FDA to include concussions, fractures, burns, temporary paralysis, and temporary loss of sight, hearing, or smell (Alder 1993).

Today, the FDA maintains a database that houses these medical device reports. Anyone involved in the development and distribution of medical devices should be familiar with **Manufacturer and User Facility Device Experience (MAUDE) database** and refer to it often. Knowledge of adverse experiences that have been reported could help to prevent future adverse events (US FDA 2020).

Design Controls

Design controls were born from this act, making the systematic review of a design an integral part of the development process. **Design control** begins with a review of user needs and intended use, leading to the development and approval of design inputs. Measurable physical and performance characteristics are used as a basis for device design. Design control includes cycles of review for the design and associated manufacturing processes, including transfer processes. The implementation of design control helps to increase transparency in the design process, providing the opportunity to recognize problems earlier, make corrections, and adjust resource allocations in the overall scheme. For example, design control should help to identify deficiencies in design input requirements, as well as discrepancies between the proposed design and those requirements.

Design control applies to new designs as well as modifications or improvements to existing device designs. It spans the life of a device, extending beyond device production through distribution, use, maintenance, and eventually, obsolescence. It applies to all changes to the device or manufacturing process design, including those occurring long after a device has been introduced to the market. This includes evolutionary changes such as performance enhancements and revolutionary changes such as corrective actions resulting from the analysis of failed product. The changes are part of a continuous, ongoing effort to design and develop a device that meets the needs of the user and/or patient. Thus, the design control process is revisited many times during the life of a product (U.S. FDA, Center for Devices and Radiologic Health 1997). The FDA has published a guidance document for design controls that can be found on the FDA website (www.fda.gov).

International Standards Organization (ISO) 13485

Countries outside the United States may have different regulations, standards, or GMPs. However, if a product made in another country is to be marketed in the United States, the manufacturer must meet U.S. GMPs, and the product must be cleared by the FDA. While each country has its own regulations and standards, in recent years, the medical device community has begun to harmonize internationally. In fact, at the same time as the SMDA was enacted, the FDA made an effort to make CGMP regulation consistent with applicable international standards to the extent possible.

The International Organization for Standardization (ISO) is an independent organization with more than a hundred different member nations (162 to date). Its experts combine their collective knowledge to develop voluntary "market relevant international standards that support innovation and provide solutions to global challenges" (International Organization for Standardization [ISO] 2018). The ISO has published international standards for almost every business sector. You can learn more about ISO on its website (https://www.iso.org/home.html).

You may have heard about the ISO 9000 family of quality-management standards and in particular ISO 9001. For many U.S. device manufacturers, implementation of this family of standards was the first effort to harmonize with international standards. However, the ISO 9001 standard did not fully serve the GMP requirements of the FDA, and there was also concern that if ISO 9001 was revised—which occurred from time to time—the FDA requirements might be even less served.

Thus, more work was done to supplement those standards specifically for medical devices. By the end of 1996, the U.S. *Federal Register* published a final ruling regarding the incorporation of a quality system in the CGMP regulations, as well as including the provision for design control. **ISO 13485** is the body the international standards adopted in that work (U.S. FDA, Center for Devices and Radiologic Health 1997). This document is a set of quality management systems specifically for medical devices that outlines requirements for regulatory purposes. It has undergone some revision, and the latest version was published in March 2016 (ISO 2018).

This standard is designed to be used by organizations involved in the design, production, installation, and servicing of medical devices and related services and even for other parties that are responsible for reviewing, auditing, or certifying a process. Certification to ISO 13485 is voluntary. Although not a requirement of the standard, many medical device manufacturers will attest that certification is very helpful in demonstrating to regulators that they have met the requirements of the standard (ISO 2018).

Summary

Today, regulatory and quality-management practices are still strongly driven by events and precedents. Although medical devices are much safer today than they were before 1906, there are still, unfortunately, situations where people have been injured or killed because of an unanticipated risk or hazard in some aspect of medical device design. Once the cause of an injury or death is associated with a medical device or some aspect of its design, practices will be modified. It is significant that the cause is not always clear-cut, but to ensure minimal risk, that device or aspect of the device will not be recommended. A device designer would be well served to learn about the history of a technology that may be considered in the design of a specific device, especially with regard to the safety of that technology.

A good example today is in the design and selection of the power supply for an electronic medical device. While many would argue that the use of a lithium ion battery may be optimal for the device—providing for the longest life and smallest sized battery—recent accounts of the explosion of lithium ion battery–powered devices, such as cell phones and electronic cigarettes, have established significant concern for the safety of these batteries in medical devices. Therefore, until more evidence can be developed to ensure the safety of these batteries, a designer might opt to use an older technology, such as a nickel-cadmium battery.

A recurring theme in medical device regulations and processes is *flexibility*. Rather than being presented as a specific list of requirements and procedures, the regulations are provided as guidelines. The interpretation of these guidelines can be a source of confusion or heated debate, even among experts in the field. Certainly, anyone who has been involved in the development of a medical device has been faced with challenging questions—and sometimes frustrations—about which design option might be most effective and also best satisfy the guidelines. But the guidance format allows for similar guidelines to be applied to many different types of devices. It also, and most important, allows for innovation, improvement, and optimization of effectiveness in device development and performance. When applied with good background evidence and communication, the guidelines should lead to the safest and most efficacious devices possible at the time of development.

Study Questions

1. Draw a timeline of the events described in this chapter, and in just a few words, note the significance of each event.

2. Why do you think there was resistance to regulations over food and drugs until the turn of the twentieth century?

3. Why was the Pure Food and Drug Act of 1906 amended in 1938? List some of the benefits of the amendments.

4. Until the Food, Drug, and Cosmetic Act of 1938 was instituted, the FDA was required to initiate action against a product after it was determined to be misbranded or adulterated. What change in regulatory power came as a result of this act?

5. Why are GMPs not a collection of detailed procedures?

6. What new reporting requirements were brought about by the Safe Medical Devices Act of 1990? Why were they needed?

7. Before 1990, FDA auditors were limited to examining medical device production and quality control records. What additional records were they able to review after enactment of the Safe Medical Device Act? How do you think this change can better ensure that a medical device is safe and effective?

Thought Questions

1. Go to the FDA website and review the history of the FDA. Add into your timeline 5 to 10 significant events since 1906, and in just a few words, describe the impact of each event (i.e., describe what regulations or practices changed as a result of the event).

2. Are maggots an approved medical device? Explain your answer.

3. Describe the relationship between the benefits and challenges of the medical device regulatory and quality structure.

References

Alder, H. C. 1993. "Safe Medical Devices Act: Management Guidance For Hospital Compliance With the New FDA Requirements." *Hospital Technology Series* 12(11):1–27.

Ballentine, Carol. 1981. "Sulfanilamide Disaster," *FDA* Consumer, June. www.fda.gov/files/about fda/published/The-Sulfanilamide-Disaster.pdf.

Barkan, I. D. 1985. "Industry Invites Regulation: The Passage of the Pure Food and Drug Act of 1906." *American Journal of Public Health* 75(1):18–26. https://doi.org/10.2105/AJPH.75.1.18.

Hyman, William A. 1992. "The Medical Device Industry: Science, Technology, and Regulation in a Competitive Environment." *Annals of Biomedical Engineering* 20(2):256–257.

International Organization for Standardization (ISO), "About ISO," 2018. www.iso.org/iso/home/about.htm.

Lee, Mark H., Patrick Au, John Hyde, et al. 2015. "Translation of Regenerative Medicine Products into the Clinic in the United States." In *Translational Regenerative Medicine*, New York: Elsevier, pp. 49–74. https://doi.org/10.1016/B978-0-12-410396-2.00005-0.

National Institutes of Health (NIH). 2018. "The Hygienic Laboratory: Abutment." National Institute of Allergy and Infectious Disease, Bethesda, MD. www.niaid.nih.gov/about/joseph-kinyoun-indispensable-man-hygienic-laboratory.

"The Pure Food Bill." 1906. *Science* 24(606):185–189. https://doi.org/10.1126/science.24.606.185.

U.S. Food and Drug Administration (FDA). 2006. "Milestones in U.S. food and drug law history." FDA, Washington, DC., pp. 1–13. https://www.fda.gov/AboutFDA/History/FOrgsHistory/EvolvingPowers/ucm2007256.htm.

———. 2020. "Manufacturer and user facility device experience." U.S. Department of Health and Human Services, Washington, DC. www.accessdata.fda.gov/scripts/cdrh/cfdocs/cfmaude/search.cfm.

U.S. Food and Drug Administration (FDA), Center for Devices and Radiologic Health. 1997. "Design Control Guidance for Medical Design Manufacturers." Silver Spring, MD. www.fda.gov/downloads/MedicalDevices/DeviceRegulationandGuidance/GuidanceDocuments/ucm070642.pdf.

U.S. Government Publishing Office. 2019. "Electronic Code of Federal Regulations." USGPO, Washington, DC. http://www.ecfr.gov/cgi-bin/text-idx?SID=eba8061a5b3919889d84c2aaa19e61fd&mc=true&node=pt21.5.314&rgn=div5#sp21.5.314.h.

CHAPTER 4

Thalidomide

Learning Objectives

Recognize how the thalidomide tragedy influenced stricter and more specific regulations when bringing a new product to market.

Gain insight into the competing objectives and dynamics between a company and a Food and Drug Administration reviewer during the new drug/product application process.

Recognize why human subject testing requires informed consent.

Develop an appreciation for the scientific rigor needed when preparing evidence based on preclinical and clinical testing.

New Terms

thalidomide

peripheral neuropathy

informed consent

adverse events

preclinical tests

enantiomer

teratogenic

chiral

Thalidomide Tragedy

As we saw in Chapter 3, events and precedents have driven many U.S. regulatory and quality-management practices. One of the most catastrophic examples was that arising from use of the drug **thalidomide**. Thalidomide was developed in Germany in the mid-1950s, and it was originally marketed as a sleep aid that, compared to barbiturates, did not pose a risk of deadly overdosing. It gained popularity among pregnant women because it also alleviated morning sickness. Tragically, it led to birth defects in more than 10,000 children throughout the world. Fetuses that were exposed to the drug during early embryonic development developed severe congenital deformities, most notoriously manifesting in the limbs, hands, and feet, resulting in such tragic outcomes as shortened limbs and malformed fingers and toes (Figure 4.1). An affected child might have received a prosthetic to help accommodate the deformities, but such devices were fairly crude (Figure 4.2). This tragedy was the pivotal case that led to significant changes to Food and Drug Administration (FDA) oversight of new drugs.

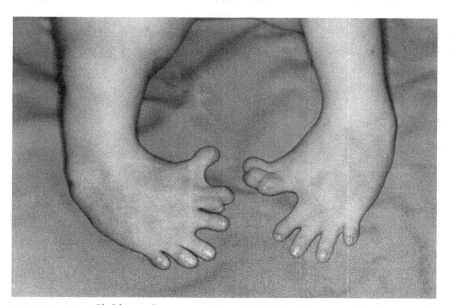

FIGURE 4.1   Children who were victims of thalidomide were often born with deformities in their limbs.

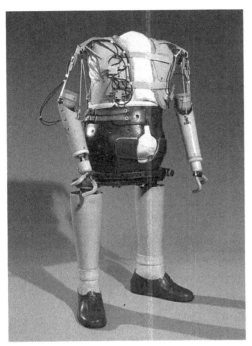

FIGURE 4.2   Artificial limbs for a thalidomide child in the early 1960s.

The Power of One FDA Reviewer

The thalidomide disaster was prevented in the United States largely because of the valor of one FDA reviewer, who asked many questions on a case that was thought to warrant a routine approval. Despite immense pressure, FDA Medical Officer Frances O. Kelsey refused to approve thalidomide for use in the United States. By 1960, the drug was approved in more than 40 countries and was gaining popularity throughout Western Europe and Canada. Still, Dr. Kelsey rejected the drug company's application with concerns that it lacked sufficient evidence for safety and that there was inadequate rigor in the clinical trials presented.

Keep in mind that while the Food, Drug, and Cosmetic Act of 1938 (FDCA) required that a new drug be tested for safety, it did not provide explicit details about how testing should be conducted, and many clinical trials were managed with minimal oversight. Dr. Kelsey felt that the representative for the drug manufacturer, Richardson-Merrell, provided evidence that was based more on physician testimonials than scientific evidence (McNeill 2017). Richardson-Merrell continued to pressure for approval, perhaps being more focused on its desire to get the drug on the market, feeling that Dr. Kelsey was overly zealous in her review. At one point the company asked that approval be granted in time for Christmas, which it said was a "busy period for sedatives" (Peritz 2014).

Ultimately, in late 1961, independent reports from Europe and Australia linked thalidomide to horrendous birth defects. By November, Germany and then Britain announced that they were pulling the drug from the market. The drug never made it to the U.S. market.

Animal Models and Pregnancy

In later interviews, Dr. Kelsey explained that despite the pressure, she felt that she had to "stick to her guns" (Peritz 2014). In December 1960, during the course of Dr. Kelsey's review, the *British Medical Journal* published reports of patients taking the drug experiencing painful tingling in their arms and feet, a sensation known as **peripheral neuropathy**. Dr. Kelsey didn't believe that such a side effect should be associated with a sleeping pill. She continued to challenge Richardson-Merrell, writing the company a letter telling it that she suspected the company knew of the toxic side effects but chose not to disclose them. Further, the reports of neurotoxicity led Dr. Kelsey to consider the safety of thalidomide when used by pregnant women, so she requested evidence to prove its safety (McNeill, 2017).

Fortunately, in her graduate research, Dr. Kelsey had state-of-the-art experience with animal models that led her to ask this key question. She studied in Dr. Eugene Geiling's laboratory at the University of Chicago. Dr. Geiling was a medical officer at the FDA, and in 1937, his laboratory was asked to investigate the deaths of 107 people caused by the anti-infective elixir of sulfonilamide (see Chapter 3). Drs. Geiling and Kelsey developed animal studies leading to the discovery that the artificial raspberry flavoring added to the elixir to hide its bitterness was made with a toxic solvent, diethylene glycol, commonly known as antifreeze (McNeill 2017).

This was a landmark case that led to the FDCA of 1938. Recall that this was about the same time that the FDCA was under review by Congress, but that review had met many hurdles and delays. The thalidomide tragedy, with so many victims being children, was the impetus needed to create a law that required drug manufacturers to show that a drug was safe before it could be marketed. Still, animal studies were not required by the FDCA of 1938, nor were they required even 20 years later, when thalidomide was introduced.

Dr. Kelsey had another key experience years later. She had remained in Dr. Geiling's laboratory after she earned her Ph.D., joining a wartime effort during World War II to find treatments for soldiers with malaria. Her work involved studying the metabolism of drugs in rabbit models. In that work, she discovered that while rabbits normally had a liver enzyme that enabled them to easily break down quinone, pregnant rabbits could not break the same drug down as readily. Further, she discovered that rabbit embryos could not break the quinone down at all (McNeill 2017). This made her think that medicine could affect a pregnant woman and her embryo differently.

It might be hard to believe today, but at that time, there was very little research being done in the field of drugs and their effect on embryos. Thankfully for the families of children who were born in the United States during the early 1960s, Dr. Kelsey had such experience and thought to consider it in her review of the thalidomide application.

Kefauver-Harris Drug Amendments Act of 1962

Meanwhile, Senator Estes Kefauver had been holding hearings on the control and quality of drug testing as it pertained to a drug's effectiveness. The hearings also addressed concerns for some extraordinary claims that drug makers were making in their labeling and advertising, as well as for the general high costs of many drugs. During those discussions, a renowned clinical pharmacologist from Johns Hopkins University spoke about the differences between well-controlled studies and the typical drug study of the period. With the awareness generated by thalidomide, Senator Kefauver and Senator Oren Harris seized the opportunity to drive a bill through Congress, the Kefauver-Harris Drug Amendments to the federal FDCA. President John Kennedy signed the bill in October 1962.

These amendments introduced some major changes to the way a new drug could be introduced to the U.S. market. The new requirements included that before marketing a drug, a company must provide evidence of effectiveness. Specifically, well-controlled studies must adequately demonstrate that the drug is effective for its intended use. Further, the FDA was given oversight to regulate the advertising of prescription drugs.

Prior to these amendments, FDA reviewers had 60 days to review the safety of a new drug application and then respond. If the reviewer did not act within that 60-day period, the drug was automatically approved, making it easy for an unproven drug to find its way into the market by this passive mechanism. In these amendments, the approval process was tightened up so that the FDA had to specifically approve the marketing application before the drug could be marketed.

Despite not yet having approval from the FDA, Richardson-Merrell representatives distributed samples of thalidomide for trial. At the time, this was permissible, and there were no rules for testing, investigation, and experimentation on humans. With the aim to protect people from unwittingly becoming the subject of unproven drug testing, the Kefauver-Harris Drug Amendments included requirements for the **informed consent** of study subjects and that **adverse events** be reported to the FDA. Companies could no longer distribute samples without enrolling someone in a clinical study that obeys these requirements.

Advancing Animal Models

You might wonder why thalidomide was so broadly approved for use in other countries. Contrary to popular belief, animal testing had been performed on thalidomide, but those tests were unable to predict the dangers of the drug. It was in part because of these results that the drug was made available to doctors.

The animal tests were very cursory, and this event is credited with advancing the requirement for animal testing in the specific protocols we see today. **Preclinical tests** were not performed in pregnant animals prior to the drug being used in patients. Shortly after thalidomide was withdrawn from the market, it was shown to cause fetal toxicity in some animals (Botting 2002). On further investigation, it was found that drug reactions and responses are different among various species; this was the first time such differences were ever recognized. It was observed that mice, most commonly used to screen for drug action, are less sensitive to thalidomide than other species, although the reason is still not clear. Because of thalidomide, drug screening policies have changed to incorporate several species of animals, pregnant animals, and in vitro tests (Vargesson 2015).

Stereoisomers and the Aftermath

We continue to study thalidomide today. It is still used to treat leprosy, multiple myeloma, and cancers, as well as Crohn disease, human imunodeficience virus (HIV) infection, and others disorders. From the painful lessons of 60 years prior, today caution is afforded to patients to ensure that they are not pregnant.

Through our studies, we learned that thalidomide has an interesting chemical property. It is produced as a stereoisomer, and only one form of the isomer is believed to be harmful. You may recall that stereoisomers can be mirror images but can never be superimposed on one another, much like your right and left hands. The $R(+)$ **enantiomer** is the desired sedative, and the $S(-)$ form is **teratogenic** (Vargesson 2015). For many years after the thalidomide disaster, scientists believed that the disaster may have been averted if countries were more careful in their study of the drug before releasing it to the public. In fact, there are many drugs available today with similar **chiral** characteristics. Based on the lessons learned from thalidomide, today's drugs are produced and distributed in pure form instead of mixed form, as thalidomide was in the 1950s.

It remains unclear whether production and distribution of only the $R(+)$ enantiomer of thalidomide would have averted the global disaster it caused. Recent studies demonstrate that both the R and S enantiomers can rapidly interconvert in body fluids and tissues and form equal concentrations of each form (Vargesson 2015). Thus, even if only the harmless enantiomer was taken, the body might convert part of it to the harmful enantiomer after it is ingested.

Important Design Knowledge

This case demonstrates a recurring theme in biomedical engineering and medicine. Biology has its own set of rules, and those rules trump all other laws of science. If biology is not obeyed, even the most innovative technology can fail, sometimes with grave consequences. While we have gained a better caution for the gravely different effects of these mirror-imaged molecules, we still cannot predict with certainty how they will perform in vivo in fetuses in pregnant women. But we are wiser today, and we are more careful about what we do not know. This is one of the most important lessons to maintain in biomedical research.

Summary

When Dr. Kelsey took on the thalidomide application, it was her second file as an FDA reviewer. She had joined the staff of three FDA reviewers only a month prior. Some reviewers worked only part time. At that time, drug manufacturers were only required to provide evidence of safety, but the degree of scientific rigor was not mandated. There was no requirement to prove efficacy. The concept of controlled trials was still in early development, and many marketed drugs were not effective for their labeled uses.

At the time, Dr. Kelsey had asked questions about safety at depths beyond what was practiced and considered reasonable by many. She did not have the support of detailed laws about efficacy or human subjects testing or guidelines for animal models and for testing the effects of a drug on embryos. Her diligence averted the thalidomide disaster in the United States. In 1962, President John F. Kennedy presented her with the President's Award for Distinguished Federal Civilian Service, the highest award the government can give to a civilian. Dr. Kelsey's actions led to many critical new drug regulations in the Kefauver-Harris Drug Amendments. To this day, the exact mechanism by which thalidomide causes birth defects is not fully understood. The research that it spawned, though, and, just as important, the review and testing practices that it inspired have left an indelible mark on today's regulatory procedures. In 2010, the FDA named a prize after Dr. Kelsey in recognition of these significant contributions.

Study Questions

1. Was thalidomide advertised for use by pregnant women? What was the drug's primary use?
2. How were U.S. citizens able to take thalidomide, even without FDA approval?
3. Why can we often not predict with certainty the behavior of a drug in the body?
4. Dr. Kelsey won praise for not bowing to pressure from the thalidomide drug developers. Why were they pressuring her? What are the competing objectives between regulators and developers?
5. Define *adverse events* as it is used in a clinical trial.

Thought Questions

1. Outline the experiences Dr. Kelsey had that made her an effective choice for reviewing thalidomide.
2. List the changes that were included in the Kefauver-Harris Drug Amendment and explain the intended impact of each. If these amendments were in place before thalidomide, how would they have helped Dr. Kelsey and FDA reviewers?
3. What does informed consent consist of? Why is informed consent required in clinical testing?
4. What was wrong with the preclinical models of thalidomide? In what way did they fail to show safety?
5. Using thalidomide as an example, explain why physiologic knowledge is so important in the development of a drug or medical device.

References

Botting, J. 2002. "The History of Thalidomide." *Drugs News and Perspectives* 9(15):604–611. https://doi.org/0.1358/dnp.2002.15.9.840066.

McNeill, L. 2017. "The Woman Who Stood Between America and a Generation of 'Thalidomide Babies': How the United States Escaped a National Tragedy in the 1960s." *Smithsonian Magazine*, May 8. www.smithsonianmag.com/science-nature/woman-who-stood-between-america-and-epidemic-birth-defects-180963165/.

Peritz, I. 2014. "Canadian Doctor Averted Disaster by Keeping Thalidomide Out of the U.S." *Globe and Mail Canada*, November 24, pp. 1–6.

Vargesson, N. 2015. "Thalidomide-Induced Teratogenesis: History and Mechanisms." *Birth Defects Research Part C: Embryo Today* 105(2):140–156. https://doi.org/10.1002/bdrc.21096.

CHAPTER 5

The Dalkon Shield

Learning Objectives

Recognize the need for the Medical Device Amendments of 1976.
Understand how these amendments affect today's medical device marketing policy in the
   United States.
Learn about intrauterine devices.
Appreciate how precedent device experiences affect revised attempts.

New Terms

preamendment status
intrauterine device (IUD)
Supramid
pelvic sepsis

Preamendment Status

You have probably heard about some medical devices "grandfathered" into the U.S. marketplace. These devices were not required to obtain approval by the Food and Drug Administration (FDA) because they were in commercial distribution before enactment of the Medical Device Amendments of 1976; in other words, they have **preamendment status**. We learned in Chapter 3 that these amendments established guidelines for evaluating a device based on its risk.

As discussed in Chapter 4, momentum was gaining toward more FDA oversight of medical devices during the 1960s. As with drugs in the earlier part of the century, the need for stricter supervision of medical devices was brought to light by events that were often tragic. The events that led to the Medical Device Amendments of 1976 were no exception. Consider the case of the Dalkon Shield.

Dalkon Shield

An **intrauterine device (IUD)** is a device placed into the uterus to prevent pregnancy. The device sits at the top of the cervix and is configured to press onto the uterine wall. Modern IUDs were introduced in the early 1900s and were made mostly of metals. They required that the cervix be dilated for insertion. Women who used them risked injury from the rigid materials and further risked infection because of the lack of antibiotics and sterile practices in that era. However, although the precise mechanism by which IUDs prevented fertilization was not well understood, they were used safely and effectively by many women for decades.

In the early 1960s, this was a topic of interest for two researchers at Johns Hopkins University, Hugh J. Davis, a gynecologist, and Irwin Lerner, an engineer. In 1966, they believed that they had discovered the mechanism by which an IUD prevented pregnancy. With that knowledge, they developed a new IUD, the Dalkon Shield.

Our social and political climates could not have been more ripe for such a development. Global population was growing explosively. Contraception practices were becoming more acceptable, and laws were changing to reflect that acceptance.

In the 1960s, the birth control pills that were available often induced serious side effects and adverse events such as strokes. The use of an IUD seemed like a reasonable alternative. The use of IUDs grew significantly, culminating in use by more than 12 million women globally by 1970. A quarter of those women resided in the United States (Centers for Disease Control and Prevention [CDC] 1982).

The rapidly emerging field of plastics made for opportunites in medical device design that were not previously available with traditional materials. Flexible, softer materials would be less likely to cause injury and bleeding compared with the rigid metals used in earlier IUDs. One problem with IUDs had been their tendency to expulse, or leave the uterus, unwantingly. This rendered the IUD ineffective. An IUD of flexible plastic could be designed to counter the potential for undesired expulsion, making it more effective. The researchers who designed the Dalkon Shield believed that the rationale for pregnancy prevention was related to a noninfectious inflammatory response from surface contact of the IUD with the endometrium. Flexible plastics made it feasible for them to design a device that maximized surface contact with the endometrium, thus, they believed, improving the effectiveness of the device.

The Dalkon Shield was a thin plastic disk with small fins on its perimeter to help lodge the device in place and to prevent it from falling out. It also had a tail string to facilitate removal (Figure 5.1). The tail string was made differently from that of other IUDs: it was a nylon, cable-like structure instead of being just a string, and it used a different material, **Supramid**. This was due to the Dalkon Shield being more difficult to remove than other IUD designs at the time.

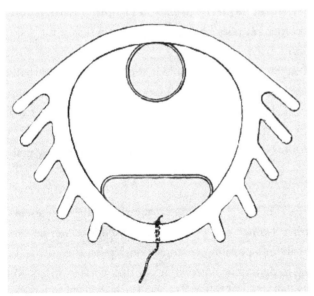

FIGURE 5.1   The Dalkon Shield included fins for lodging and a thick polymer tail string to facilitate removal.

Rogue Clinical Studies

In 1968, Dr. Davis, performed a 12-month study involving 640 women from Baltimore and published very encouraging results (Roepke and Schaff 2014). His claims included low pregnancy and expulsion rates, high retention rates, and reduced bleeding complications. With these study results, the Dalkon Shield was agressively marketed, and by 1974, nearly 4 million devices were sold; about half were sold in the United States.

Not much was known about how IUDs worked and how to handle complications. The *inflammation theory* promoted by Dalkon Shield researchers turned out to be inacurate. Today, we know that IUDs prevent pregnancy by inducing a thickening of cervical mucus and/or by preventing ovulation. An IUD may also include a synthetic hormone to further reduce the likelihood of pregnacy.

The misconceptions in the 1960s and 1970s about how IUDs prevented pregnancy sometimes had grave consequences. For instance, in cases where a woman did become pregnant even with an IUD inserted, the common advice was to leave the IUD in. It was suggested that the fetus would push the IUD aside as it grew. What we learned, unfortunately, was that this can cause a lot of harm both to the mother and to the fetus.

When reports of problems began to arise from women who used the Dalkon Shield, the company that made the device, A.H. Robins, simply responded with clinical work-arounds, such as adding an anesthetic regimine to mitigate painful insertion and removal. Or the company blamed the women for poor hygeine and their physicians for poor insertion technique if complications arose.

Still, concerns for Dalkon Shield users prevailed, the greatest being that of pelvic infection. Other complications included uterine perforations and septic abortions (a septic abortion occurs when the placenta becomes infected and causes embryo death). In many cases these complications would kill the IUD user too.

In 1975, the *Journal of the American Medical Association* published a study that reviewed these concerns. It stated:

> The recent report of 209 cases of septic spontaneous abortion and 11 maternal
> deaths in the United States in women using the Dalkon Shield intrauterine device
> (IUD) raised the question about a possible causal relationship between the IUD and
> **pelvic sepsis.** It is essential to determine whether or not this sepsis is unique to the
> Dalkon Shield or generic to all types of IUDs. Our studies permit the conclusion
> that the tail of the Dalkon Shield is structurally and functionally different from
> the tails of the four other IUDs tested. The unique characteristics of the Dalkon

tail theoretically could provide a mechanism whereby pathogenic bacteria from the vagina enter the uterine cavity and cause sepsis [Tatum et al. 1975].

Momentum Builds to Expose the Truth

Lawsuits began to emerge. Initially, the company retaliated against lawsuit threats with unlimited resources. The company prolonged cases and applied tactics to wear out the limited resources of individual claimants (Roepke and Schaff 2014). But there were so many claims that attorneys were able to collaborate unlike ever before. They shared information to help build their claimants' cases. They rallied public scrutiny, maintaining pressure on the company.

In December 1973, the first of several hundred lawsuits came to trial. A woman named Connie Deemer had become pregnant while using the Dalkon Shield. She later suffered a perforated uterus. She had to undergo surgery for her life-threatening condition. In his investigation, her attorney discovered that Dr. Davis had a financial stake in the Dalkon Shield. Dr. Davis had previously testified before the U.S. Senate, denying any financial interest (Bankston 2003). The prosecutor also found evidence that the company had withheld or destroyed negative information about the Dalkon Shield, such as birth rates being as much as five times higher than published, knowledge that the tail string caused infections, and product changes that were not tested. His research revealed that a Robins employee who had brought the tail string issue to the attention of management had been fired.

During this same trial, it was discovered that the company had deliberately deceived the FDA, claiming that it added copper only to improve the capability of imaging. However, the company actually included it to improve the spermicidal properties of the contraceptive. Recall that the FDA did not regulate medical devices when the Dalkon Shield was introduced to the market in the early 1970s. Had the intended use of copper been properly identifed, the FDA would have been authorized to regulate it based on the drug oversights that were in place and would have then been able to require thorough testing. A paper reporting on the inhibitive action of copper on fertilization had only recently been published (Margulies 1975). But the effects had not been well understood or widely reviewed at the time.

Finally exposed, A.H. Robins could no longer fight back. More details of the company's wrongdoings were brought to light. It became clear that the study that Davis had performed in 1968 was poorly designed. The initial number of 640 subjects was too low. Many women had dropped out of the study after only a few months. Others were on supplemental birth control methods during the study. Subjects' medical histories of infection or sexually transmitted diseases were not monitored. The product that was tested in the study was changed before it went to market.

The FDA asked Robins to remove the Dalkon Shield from the market in 1974 and to notify doctors to stop inserting devices and also to remove any that were still in use. Although Robins agreed to stop selling the device in the United States, it continued to sell it overseas for nearly another year, and the company did not comply with the requested notification to doctors until 1980, after many more injuries and deaths occurred.

Medical Devices Amendments of 1976

Recall from Chapter 3 that the Cooper Committee was commissioned by President Nixon in 1970 to study the adverse effects of medical devices for human use. While the Cooper Committee recommendations were being debated in Congress, pacemaker failures were being reported. Congressional hearings were held in 1975 regarding the concerns that had been reported with the Dalkon Shield. The problems presented by these two devices highlighted the need for the more FDA oversight of medical devices, and the Medical Device Amendments were enacted in 1976 (Rados 2006).

Partial Justice 20 Years Later

Nearly 200,000 lawsuits were ultimately filed against the A.H. Robins Company. By 1987, the company was ordered by the courts to set up a $2.5 billion trust fund for women who were harmed by the Dalkon Shield. To receive compensation, any claimant had to produce her medical records and provide detailed descriptions of personal hygiene and sexual activity. This was embarrassing for many and impossible for others. Many women were treated in public clinics and, by this time, no longer had access to records that had been discarded. In other countries, many women were unable to produce accurate records and so did not receive compensation (Jones 1985).

A.H. Robins skirted larger payouts by declaring bankruptcy, despite its financial health at the time. Whereas $2.5 billion may sound like a lot of money, on distribution, women who suffered serious injuries from the Dalkon Shield were paid very little. Most women were paid less than $1,000. Two years later, after recovering significant corporate value, A.H. Robins was sold to American Home Products in 1989. The corporate officers who led A.H. Robins through the cover-up profited nicely in the transaction.

Aftermath of the Dalkon Shield

In the aftermath of the Dalkon Shield lawsuits, use of IUDs diminished sharply. By the early 1980s, nearly all IUDs had been removed from the U.S. market. Today, only about 5 percent of women in the United States use an IUD, despite developments and evidence that indicate better safety and effectiveness than the Dalkon Shield. Only a few options are available on the U.S. market today (Daniels et al. 2018).

Important Design Knowledge

Fifty years later, the issues associated with the Dalkon Shield still resonate with many clinicians, and this has surely affected market acceptance of new IUDs. This is not uncommon. When there is precedence of a bad experience with a medical device, it is very difficult to overcome the associated stigma.

You may be able to offer strong technical rationale for improvements that would avoid the cause of failure or injury in the precedent device. In the end, though, you can't be 100 percent certain unless you develop thorough and compelling clinical evidence. If a similar device has failed or caused harm in the past, you can be certain that there will be tremendous scrutiny on the part of the clinicians, the public, regulatory bodies, and payers (insurance companies). You may be required to develop mounds of evidence to assure everyone that the same mistakes will not happen again. You need to weigh the time and high costs involved in developing evidence as well as slow market adoption in your business plan. Often designers determine that going down that same road a second time is not justified because of these potential impediments.

Summary

The Dalkon Shield was an unsafe medical device. Its use became a pivotal case that led to stronger authority for the FDA. The business case for a new medical device—or component of that device—may be adversely affected by its history. Those who remember or who have studied this case are cautious about adopting a new IUD.

Study Questions

1. What does it mean for a medical device to be grandfathered for FDA approval?
2. How did the introduction of plastic materials offer opportunities for improving the design of the IUD?
3. List three unethical practices that were revealed during the trial for Connie Deemer.
4. If the Dalkon Shield were introduced after the Medical Device Amendments of 1976, what would the manufacturer have had to have done differently in order to be marketed in the United States?
5. Did the FDA order A.H. Robins to remove the Dalkon Shield from the market?
6. The study that Davis performed was poorly designed. Select one of the design flaws and explain why it provided misleading evidence.
7. Why were many women unable to receive compensation from A.H. Robbins' $2.5 billion trust fund?

Thought Questions

1. Describe the mechanisms by which an IUD prevents pregnancy.
2. How do you think the Dalkon Shield experience influences doctors and women when making a decision about use of an IUD today? How do you think it affects the introduction of new IUDs to the U.S. market?
3. Do you think the compensation that women received from the A.H. Robbins trust fund was fair?
4. If you or a loved one were seeking to prevent pregnancy today, what device options are available, and how do they differ in terms of safety and efficacy from those in the past?

References

Bankston, Carl L., III (ed.). 2003. *Great Events from History: Modern Scandals*, Volume 1, Salem Press.

Centers for Disease Control and Prevention (CDC). 1982. "Trends in Contraceptive Practice," 1965–1976, https://eric.ed.gov/?id=ED292000.

Daniels, Kimberly, and Joyce C. Abma. 2018. "Current Contraceptive Status Among Women Aged 15–49." *National Center for Health Statistics Data Briefs No. 327*: 2015–2017, https://www.cdc.gov/nchs/data/databriefs/db327-h.pdf.

Jones, E. L. 1985. "Dalkon Shield." *New Zealand Medical Journal* 98(783):609.

Margulies, L. 1975. "History of Intrauterine Devices." *Bulletin of the New York Academy of Medicine: Journal of Urban Health* 51(5):662–667.

Rados, Carol. 2006. "Medical Device and Radiological Health Regulations Come of Age." *FDA Consumer Magazine*, January 2006, https://www.fda.gov/files/about%20fda/published/Medical-Device-and-Radiological-Health-Regulations-Come-of-Age.pdf.

Roepke, Clare L., and Eric A. Schaff. 2014. "Long Tail Strings: Impact of the Dalkon Shield 40 Years Later." *Open Journal of Obstetrics and Gynecology* 4(16):996–1005. https://doi.org/10.4236/ojog.2014.416140.

Tatum, Howard J., Frederick H. Schmidt, David Phillips, et al. 1975. "The Dalkon Shield Controversy: Structural and Bacteriological Studies of IUD Tails." *Journal of the American Medical Association* 231(7):711–717. https://doi.org/10.1001/jama.1975.03240190015009.

CHAPTER 6

Understanding Today's Food and Drug Administration

Learning Objectives

Understand the mission of the Food and Drug Administration (FDA).

Appreciate the risk-benefit position of FDA regulations.

Gain a sense of time and costs associated with a regulated device or drug and why these factors must be considered in the design strategy.

Understand the roles of pertinent offices/divisions of the FDA and strategically compare them with regard to medical device review procedures.

New Terms

Federal Anti-Tampering Act

Physician's Desk Reference (PDR)

Class I, II, and III devices

CDER, CBER, and CDRH

combination product

FDA Mission

If you want to know how to get a new medical device approved by the FDA, you can refer to the FDA website (www.fda.gov), which has a lot of important and helpful information. If you are a novice who wants to understand how to use the information to help you develop a regulatory strategy for a new product, however, you might find all that information to be a bit overwhelming. So let's distill that down into some foundation for you. With this foundation, you should feel a lot more comfortable navigating the FDA website and related information.

From earlier chapters, we have learned that many of the regulations that are in place today are based on lessons learned, some of which had horrific consequences. From the perspective of a new product designer, getting a new product cleared for market in the United States can be viewed as exceptionally challenging, with voluminous requirements. It requires lots of detailed documentation, testing, and evidence, all of which are very expensive and time consuming to prepare. Still, as we have seen, those requirements are in place for a good reason. Your responsibilities as a designer of medical devices should align with the FDA's mission, and your job is to be as efficient as possible in meeting all regulatory requirements. Nowhere on the website or anywhere else will you find a standard and specific procedure for obtaining regulatory approval; every case is different. But if you develop an appreciation for the FDA's mission and for its mechanisms of regulation to support that mission, you will be able to anticipate many of the regulatory requirements as you prepare your product-development plan.

From the very humble beginnings described in earlier chapters, the FDA now has a huge range of responsibilities. The agency is responsible for regulating foods, drugs, biologics, medical devices, cosmetics, alcohol, tobacco, and even veterinary drugs in the U.S. market. In 2018, the FDA employed approximately 17,500 full time employees! The agency clears around 3,000 medical devices and almost 50 novel drugs each year (Woodcock 2017). A recent organizational chart for the FDA is shown in Figure 6.1.

The first part of the FDA's mission is to protect public health in a few different ways: the agency makes sure that products are safe, effective, and secure. You may ask, "What is the difference between safe and secure?" because in our minds they might mean the same thing. Safety refers to the danger inherent in the product itself. If you take a drug, for example, will that drug make you sick? Will that drug kill you? What if you take too much, that is, exceed the prescribed dosage? These types of questions and regulations to address these concerns speak to safety.

The secure portion of the mission refers to security measures used to ensure that the drug or medical device that is delivered to you is safe and of high quality. Regulations are in place to safeguard against counterfeit, stolen, contaminated, or otherwise harmful products.

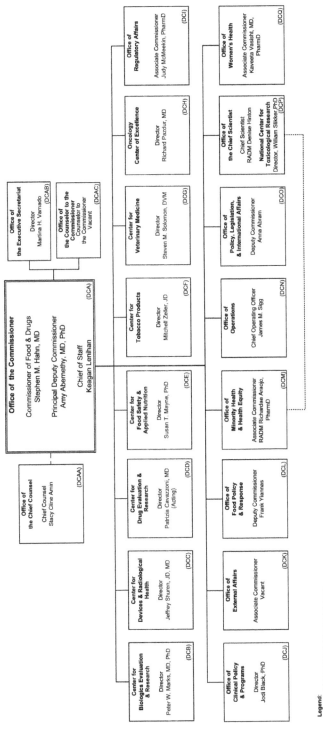

FIGURE 6.1 FDA organizational chart and description of overall responsibilities

Source: FDZ 2019b.

This oversight is in place from development through shipment along the entire supply chain, including at the manufacturer's site; through custody transfers, repackagers, and other intermediaries; and then all the way to your drug store.

Tylenol

You may not be old enough to remember, but not too long ago, many over-the-counter drugs were stored on pharmacy shelves in simple screw-top bottles. Many people who are still alive today can recall going to the drug store to buy aspirin that was placed on the shelf in a brown bottle with a simple metal screw-top cap. These bottles were not sealed or even in boxes, and anyone could open the screw cap to see bare, individual pills beneath the cotton balls that were placed there to help keep the pills from breaking in the bottle during transport. This was common business practice.

Then, in Chicago in 1982, six adults and one 12-year-old girl died abruptly for what seemed to be no reason. At first it was not clear what caused the deaths, and initially the deaths were viewed as independent incidents. Within a few days, however, the death toll mounted, and through police investigations, it became evident that all the victims had one thing in common: each had taken Tylenol capsules shortly before their death.

Police investigated and tested the pills, which they determined were laced with potassium cyanide at a level toxic enough to provide thousands of fatal doses. The investigation also showed that the pills had come from different production facilities and were sold in different drug stores in the Chicago area. Their conclusion was that someone was most likely tampering with the drug on the store shelves (Fletcher 2009).

Back then, Johnson & Johnson, the manufacturer of Tylenol, was most well known for its consumer products. It had established its reputation in American households as a trustworthy company that made gentle and safe products to take care of your family. The company had a common, trusted household name because of its Johnson's Baby Powder, No More Tears Baby Shampoo, Band-Aids, and the like.

Mass panic ensued once the link was established between Tylenol capsules and the Chicago murders. In the wake of the incident, Johnson & Johnson decided to recall and destroy all Tylenol capsules and promised that capsules would not return to the shelves until a tamperproof package had been perfected. There were an estimated 31 million bottles in circulation, and Johnson & Johnson spent millions recalling them all (Fletcher 2009).

In less than a year, with an investment of more than $100 million, Johnson & Johnson's Tylenol capsules returned to shelves in new tamperproof packaging, and market share for Tylenol was restored to the same healthy levels it enjoyed prior to the Chicago murders. The company's method of handling the recall has become a classic business case study.

As a result of this event, in October 1983, the U.S. Congress passed the **Federal Anti-Tampering Act**. Known as the "Tylenol Bill," it declared tampering with consumer products a federal offense (U.S. Food and Drug Administration [FDA] 2017).

This event significantly changed the way drugs and medical devices are packaged. Today, that same aspirin product mentioned earlier is inside a bottle that is sealed multiple times, has a childproof cap, is often packaged within a box, and that box is even sometimes wrapped in plastic. No longer could anyone conceive of taking bottles from the shelf, lacing pills with poison, and returning them to the shelf unnoticed. The Tylenol murders of 1982 exposed the former, simple packaging practices as being vulnerable to major breaches in security.

By 1989, federal guidelines were established for manufacturers to make all such products tamperproof. The guidelines aim to ensure the safety of the product you receive—in other words, protect your security. Current regulations for tamperproof packaging may be found on the FDA website (FDA 2016).

FDA Encourages Innovation

Another part of the FDA's mission is to advance the public's health. This can be done in a number of ways. The FDA encourages special innovations that make products safer, more effective, or more affordable.

You may wonder how a product could be made safer, even though it is already approved by the FDA. Well, you must keep in mind that products are approved for the U.S. market based on a very rigorous and intelligent risk-benefit analysis. If you speak to people who have undergone chemotherapy treatment for cancer, most will probably tell you that the treatment made them very sick at times, but it saved their lives. This might be an extreme example of a risk-benefit tradeoff. But just turn on your television and watch a drug commercial. Then listen closely to the legal disclaimers (the fast speaking voice-over near the end) that describe how some people may experience terrible side effects or even life-threatening adverse events (e.g., bleeding or high fever) that sometimes sound worse that the condition the drug is meant to treat! In these cases, if you were a potential candidate for the therapy, both you and your doctor would need to evaluate the risks and benefits of that treatment.

Using these same examples, imagine if you could develop a therapy—a device or a drug—that has fewer side effects, less risk of adverse events, or a higher probability of being effective compared with currently used therapies. Perhaps you could develop a cancer treatment that does not compromise the immune system as much as chemotherapy. This would be an awesome way to advance public health!

Unfortunately, some people cannot afford medical treatment. Many drugs and surgical devices are very expensive, and the overall cost of healthcare in the United States will be greater than what we can afford to spend if no cost-saving measures are introduced. If you could develop a less expensive treatment than is available today, that would certainly also advance public health.

Technology has enabled the medical device field to grow rapidly in size and sophistication over the last 50 years. There are many wonderful devices that have helped to save numerous lives. But there are many areas for improvement. As much as it is important to demonstrate your technology's safety and effectiveness and that your supply chain is secure, it is just as important to show how your technology offers improvements over currently marketed technologies. In this respect, the FDA mission likely lines up with your professional objectives to make devices better—make them safer, more effective, and less costly. As the next generation of device developers, this is your charge!

Accurate Info/Labeling

The FDA also works to ensure that patients have access to accurate, scientifically based information (FDA 2018). In Chapter 5, you learned about how FDA regulations evolved to ensure that products are accurately represented. The agency has developed procedures to prevent false advertising and claims that misrepresent effectiveness. There are also requirements to ensure that any ingredients in the product that may be harmful are clearly identified.

Direct-to-consumer advertising has been legal since 1985. Before then, the public was only able to obtain information from their doctors or from the product package inserts directly. Doctors depended heavily on information they obtained from their professional conferences, colleagues, sales representatives, and the ***Physician's Desk Reference* (PDR)**. The *PDR* lists every drug that has been approved, its indication for use, what it looks like, dosages, side effects, and adverse events. At that time, the *PDR* was not available to the general public.

Today, you can turn on the TV or get just about all you want to know from the internet. The FDA regulates to ensure that all that information is valid, and while protecting the public, the FDA has also worked to ensure that the public has more access to that information, thereby advancing public health.

Cost and Time to Market

How much do you think it costs a drug company to bring a drug to market? According to a study done in 2013 by the Tufts Center for the Study of Drug Development, the cost of bringing a new drug to market is $2.6 billion. This includes $1.4 billion of out-of-pocket costs and another $1.2 billion in lost opportunity costs. Add another $312 million dollars for research and development after approval to test new indications, formulations, dosages, and monitoring for long-term effects (Sullivan 2019).

The cost to bring a medical device to market may be quite high as well but is typically significantly lower than that for a drug. Such costs can be as low as a few hundred thousand dollars for a low-risk **Class I or II device** and can reach many millions for a high-risk **Class III device**. The cost for some devices can even exceed $100 million, especially if a lot of clinical and/or surgical data are required. Risk and classification will be discussed further in Chapter 7.

Not all drugs cost billions to bring to market. If you look at the average cost, it is around $350 million and ranges widely. But when you consider the opportunity costs, you must consider the rate of failure. Do you know what the failure rate is? If you worked in a large pharmaceutical company that had 100 drugs in the development pipeline, only 12 would make it to market. This is approximately an 88 percent failure rate (Sullivan 2019).

Medical devices tend to have a higher success rate, and there is a simpler process to bring certain types of medical devices to market. This will be reviewed in Chapter 7. But recognizing these costs should give you an appreciation for the complexity of the business and for the stakes at hand when products that require FDA regulation are developed.

To further appreciate the stakes at hand, consider the time it takes to get a new drug to market. How long do you think it takes? For an experimental drug, it takes about 12 years! For a medical device, the timeline is less demanding, with an average of three to seven years (Van Norman 2016).

This explains why many really good ideas or products never make it to market. Ultimately, these investments in dollars, time, and effort need to be justified. If a company does not believe that it will get a big enough return on its investment, it will not pursue the development of that product. Sometimes this position is not well understood at the beginning of the development process, but as the development team prepares more data about efficacy, costs, and ultimate market opportunity, a determination to proceed or "kill" will be made. These factors are continuously monitored throughout the development process and are formally reviewed at appropriate stages.

Important Design Knowledge

As a device designer, you must include the values of your resources in your decision process. Determine what pathway might cost the least or identify ways to shorten the time to bring the product to market. These decisions may be technical in nature, or they could be a business decision. Usually, though, they involve an integrated, multidisciplinary approach. While you will work alongside experts in science, the clinic, and marketing, for example, you still must be able to recognize some of the overarching challenges and opportunities in all areas in order to most effectively streamline your development efforts.

FDA Structure

For further insight, let's look at the part of the FDA that you will interact with. Within the Division of Medical Products and Tobacco, there are three offices of interest: the Center for Biologics Evaluation and Research (CBER), the Center for Drug Evaluation and Research (CDER), and the Center for Devices and Radiologic Health (CDRH). Each of these offices operates independently.

The CDER is commonly referred to as "See-der," The CBER similarly is referred to as "See-ber," And the CDRH ironically is simply referred to as "CDRH." When you seek to get your medical product approved, you will work directly with one of these three offices. You will select the appropriate office based on whether your product is a drug, biologic, or device.

Combination Products

While this may seem straightforward, there will be more and more cases of technologies that could apply to either of two different offices. There are products such as a drug-coated stent, for example; this is both a drug and a device. This is referred to as a **combination product**. You are now familiar with the costs and time required to get each type of product to market, as outlined in Table 6.1. Which office would you select if you had a choice? Of course, keeping in mind your responsibility to minimize cost and time to market, you would select CDRH!

TABLE 6.1   Relative Cost and Time to Market Among Products Approved Through Three Divisions of the FDA

Division of Responsibilities		
CDER	**CBER**	**CDH**
Center for Drug Evaluation and Research	Center for Biologics Evaluation and Research	Center for Devices and Radiological Health
"Drugs"	"Biologics"	"Devices"

← ——— →

Higher $$$, average 12 years   Lower $$, average 3–7 years

Drug-Coated Stent

If someone had not already determined guidelines, you would have to present a case to the FDA and get the agency to agree with you. This is exactly what Johnson & Johnson did as it was preparing to launch the Cypher drug-coated stent. Despite being rivals, Boston Scientific supported the same effort because it had the Taxus drug-coated stent in its pipeline. Both companies agreed that the preferred office for review would be the CDRH, given that the costs and timeline for development would likely be significantly lower.

The argument presented was basically that the office should be selected based on the primary function of the technology being presented. Because this was a stent with "just a little drug on top," it should be reviewed by the CDRH office. It should be noted that the Taxus drug was already on the market but in a different form and for a different use. The companies won their argument. You can bet that they enlisted smart lawyers, engineers, doctors, and researchers to support their case. So the convention for combination products has been to work with the office for which the product's primary function would be aimed. As in the case of a drug-coated stent, the primary function is that of a stent, so the office would be that for devices, the CDRH. As more combination products arise, the guidelines will likely evolve, and if you yourself are working with one of these types of products, be sure you are up to date on all the recent regulatory developments. You can find guidelines for combination products on the FDA website (FDA 2019a).

Why CDRH Oversees Medical Devices: X-Ray, One of the First Sophisticated Medical Devices

With so many medical devices available today, you might wonder why devices and radiologic health are covered by the same office, the CDRH. If you think about the history of devices and how they evolved, remember that devices were pretty simple at the turn of the twentieth century and then got a little more sophisticated from the 1920s through the 1950s. One of the earliest "more sophisticated" medical devices was an x-ray machine. When x-ray machines were first used, the deleterious effects of radiation were not well understood. To appreciate this, let's look at an example of an early use of x-rays.

Shoe-Fitting Fluoroscope

In 1927, inventor Jacob Lowe received a patent on his device, "Method and Means for Visually Determining the Fit of Footwear." This was a device that was placed in shoe stores that used x-rays to see how your foot fit into your shoe. In some ways akin to the current-day

Dr. Scholl's Custom Fit Kiosk for foot orthotics, it included a platform for you to stand on so that you could "visualize" the best fit for your foot. Both you and the salesperson could view the bones in your foot to see how they fit inside the shoe.

This is a claim from the 1927 patent:

The object of this invention is the effecting of means whereby the salesman, the purchaser and even the purchaser's advisory friend can visually know exactly how well a shoe is fitting, both under pressure and otherwise. To this end I have so disposed [the] X-ray apparatus as to enable the positions of the bones of the foot within the new shoe to be clearly seen and distinguished from the outline of the shoe, whereby any cramping or compression of the toes is instantly made apparent [Lowe 1927].

Imagine being a kid in the mid-twentieth century, going to the shoe store with your Mom. The salesperson asks you to try on a pair of shoes and then step up on a platform where you can stick your feet under a fluoroscope and observe from the sight glass above. Two additional sight glasses were included so that your Mom and your salesperson also could view the fitting (Figure 6.2).

FIGURE 6.2   A typical shoe store fluoroscope from the mid-twentieth century.

While the Dr. Scholl's kiosk is completely safe, the shoe-fitting fluoroscope emitted dangerous levels of x-rays, even in one simple application. Imagine just how much radiation this exposed store workers to over a period of time!

By contrast, think about how cautious we are today with far lower levels of radiation exposure. During your career, you may work with products that are used in surgery. If so, you may find that you need to observe many surgeries. For example, if you are working with an orthopedic surgical product, you may have to observe a surgery for training and product development. Before you can go into the surgical suite, however, you will need to wear about 10 or more pounds of lead! You will need to wear a vest lined with lead and maybe even a collar lined with lead to protect your thyroid. The amount of x-ray exposure that the shoe-fitting fluoroscope gave to shoe customers and store workers would be unthinkable by today's standards.

Thankfully, by the late 1950s, the shoe-fitting fluoroscopes fell out of favor. The world was beginning to learn more about how harmful radiation is. We had suffered through World War II, which introduced nuclear weapons, and the longer-term effects of radiation exposure were becoming better understood as a result of the suffering endured by the people who were exposed to that radiation. This is just one more example of how a catastrophic event shaped our regulatory policies of today.

By the way, the shoe-fitting fluoroscope was never classified as a medical device. In fact, most radiation-emitting products are not considered to be medical devices. Television receivers, tanning beds, and microwave ovens are all products that emit radiation but are not medical devices. Still, these devices must be regulated for safe radiation emissions. However, if a device makes any medical claims, the product is a medical device that must meet the regulations for both medical devices and radiation-emitting products. The CDRH has the expertise to review all radiation-emitting devices, and its expertise in medical devices grew from the introduction of such devices.

The CBER and Medical Devices

We haven't spoken much about the CBER. Presently, it is less likely that a medical device designer will interact with that office. There is one exception to note, however. Because the CBER has expertise in blood, blood products, and cellular therapies, all medical devices associated with blood collection and processing procedures, as well as those associated with cellular therapies, are reviewed by the CBER. The CBER also has expertise in the integral association of certain medical devices with these biological products and will therefore apply its review and will also ensure that all medical device laws and regulations are applied.

Summary

The FDA has a huge responsibility to protect and advance public health. This drives the agency to find ways to weigh the risks and benefits of any regulated product. These products include drugs, food, cosmetics, and medical devices. The FDA has divisions that specialize in the evaluation of regulated products, though there may be some overlap as technology advances. In these cases, a careful strategy is needed to find the most expedient pathway forward. Nonetheless, the potential for time and dollar costs must be considered in the development of a regulated product. The hurdles presented in terms of time and cost can be frustrating, but they are necessary to balance the risks associated with new medical products. You must keep in mind that for this reason, not all good technical ideas will be good business ideas.

Study Questions

1. Why was the Federal Anti-Tampering Act passed? How does this relate to the mission of the FDA?
2. What is the average success rate for a drug in the development pipeline?
3. How long does it take, on average, for a medical device to be approved?
4. What are the three offices of interest a medical device designer would potentially work with? What division are they housed under?
5. What is a combination product? What makes the regulation process for combination products unique?

Thought Questions

1. Was it necessary for Johnson & Johnson to recall all Tylenol bottles? Why was this an effective strategy?
2. Part of the FDA's mission is to ensure accurate public health information. What challenges and opportunities do you think the agency faces in the modern age? As an example, think of the way that pharmaceuticals are marketed in the United States.
3. What office would you contact if you were designing a medical filter to mitigate blood infections?
4. What is the average cost of bringing a medical device to market? What impacts do you think this high cost has on what medical devices get produced and what don't?
5. The FDA was formed before information technology/the internet and medicine were as tightly intertwined as they are now. What problems or opportunities do you see for regulatory requirements in products moving forward?

References

Fletcher, Dan. 2009. "A Brief History of the Tylenol Poisonings." *Time*, February 9.

Lowe, Jacob J. 1927. "Method and Means for Visually Determining the Fit of Footwear." U.S. Patent No. 1614998A.

Sullivan, Thomas. 2019. "A Tough Road: Cost to Develop One New Drug Is $2.6 Billion; Approval Rate for Drugs Entering Clinical Development Is Less Than 12 Percent." *Policy and Medicine*, March 2019. www.policymed.com/2014/12/a-tough-road-cost-to-develop-one-new-drug-is-26-billion-approval-rate-for-drugs-entering-clinical-de.html.

U.S. Food and Drug Administration (FDA). 2016. 21 CFR Part 884. *Federal Register*, vol. 81.

———. 2017. "Investigations Operations Manual: Federal Anti-Tampering Act." FDA, Washington, DC. www.fda.gov/media/75496/download.

———. 2018. "What We Do." FDA, Washington, DC. www.fda.gov/aboutfda/whatwedo/default.htm.

———. 2019a. "Combination Products." FDA, washington, DC. www.fda.gov/combination-products/about-combination-products.

———. 2019b. "FDA Organization Overview." FDA, Washington, DC. www.fda.gov/about-fda/fda-organization-charts/fda-organization-overview.

Van Norman, Gail A. 2016. "Drugs, Devices, and the FDA: 2. An Overview of Approval Processes—FDA Approval of Medical Devices." *Journal of the American College of Cardiology: Basic to Translational Science* 1(4):277–287. https://doi.org/10.1016/j.jacbts.2016.03.009.

Woodcock, Janet. 2017. "2017 New Drug Therapy Approvals." FDA, Washington, DC. www.fda.gov/downloads/AboutFDA/CentersOffices/OfficeofMedicalProductsandTobacco/CDER/ReportsBudgets/UCM591976.pdf.

CHAPTER 7

Preparing a Regulatory Strategy

Learning Objectives

Differentiate elements and procedures associated with each medical device classification.

Appreciate the nature, value, and risk associated with a preamendment device.

Recognize how the value proposition for a product changes between the different device classifications.

Understand what a 510(k) is.

Grasp the foundation and rationale for common practices that are applied to streamline the process for medical device use and regulatory approval.

New Terms

intended use

indications for use

general controls

special controls

exempt

preamendment device

Code of Federal Regulations (CFR)

premarket notification (PMN)

510(k)

premarket approval application (PMA)

sponsor

investigational device exemption (IDE)

institutional review board (IRB)

humanitarian use device (HUD)

nonsignificant risk (NSR)

substantially equivalent

predicate

de novo

outside the United States (OUS)

CE mark

FDA's Rationale for Medical Device Classification

As discussed previously, before a product gets to market, it has to meet the regulatory requirements for that device. In most cases, this includes testing so that efficacy and safety are proven. Your product-development team will need to determine which tests to perform and what types of data need to be prepared and evaluated by the Food and Drug Administration (FDA). For a medical device, the type of evidence required depends on the **intended use** and **indications for use**, as well as other marketing requirements. Intended use means the general purpose of the device; it describes the function of the device and includes the indications for use. Indications for use are more specific and describe "the disease or condition the device will diagnose, treat, prevent, cure or mitigate" and includes a description of "the patient population for which the device is intended" (U.S. Department of Health and Human Services et al., 2014).

You will need to suggest the appropriate class for your device. The classification, in turn, will dictate the regulatory pathway that you must follow. The FDA offers guidelines for these determinations. It is important to develop an appreciation for the classification system. This will help you to interpret the guidelines, especially when you are introducing a novel new product. The FDA has the final say regarding device classification, but as you will learn, you must first prepare evidence for your suggested classification and include that evidence in your application before the FDA issues a determination.

As indicated in Chapter 6, there are three classes of medical devices. The classes are divided based on the degree of control needed to ensure safety and efficacy. That control is determined by the amount of risk associated with the device. Table 7.1 summarizes the rationale for medical device classification.

General Controls

General controls apply to all medical devices unless exempt (see the section "Exempt"). These regulatory requirements have been authorized by the Food, Drug, and Cosmetic Act (FDCA) of 1938 (see Chapter 3). Table 7.2 lists the types of general controls and the sections of the FDCA in which those controls are discussed.

Special Controls

Special controls are usually device specific. If general controls alone are insufficient to provide reasonable assurance of the safety and effectiveness of the device, then the device

TABLE 7.1   Medical Device Classification Based on Relative Risk

Risk	Class	Degree of Control Needed to Assure Safe and Effective	Example	% of Medical Devices on Market
Low to moderate	I	General controls	Gauze Bedpan Wheelchair	47%
Moderate to high	II	General conrols and special controls	Needles Powered wheelchair	43%
High	III	General controls and PMA	Pacemaker Breast implant	10%

Note: Class I devices pose low to moderate risk to the patient or end user. Class II devices pose moderate to high risk. Class III devices "(1) support or sustain human life, (2) prevent impairment of human health, (3) may present a potential unreasonable risk of illness or injury where general controls and special controls are insufficient to provide reasonable assurance of the safety and effectiveness of a device, or (4) if there is insufficient information to make such a determination."

Source: U.S. FDA 2018a.

TABLE 7.2   General Controls Applied to All Medical Devices Unless They Are Exempt

501	Adulterated devices
502	Misbranded devices
510	Registration of producers of devices: • Establishment registration and device listing • Premarket notification (510k) • Reprocessed single-use devices
516	Banned devices
518	Notifications and other remedies: • Notification • Repair • Refund • Reimbursement • Mandatory recall
519	Records and reports on devices: • Adverse event report • Device tracking • Unique device identification system • Reports of removals and corrections
510	General provisions respecting control of devices intended for human use: • Custom device • Restricted device • Good manufacturing practice requirements • Exemptions for devices for investigational use • Transitional provisions for devices considered as new drugs • Humanitarian device exemption

Note: This table describes in which section of the FD&C Act these General Controls are discussed.

Source: U.S. FDA 2018a.

may be classified into Class II *if* there is sufficient information to establish special controls to provide that assurance. Table 7.3 provides a list of special controls.

TABLE 7.3   Any or All of These Special Controls May Be Required in Addition to General Controls for a Class II Device

Performance standards	Patient registries	Premarket data requirements
Postmarket surveillance	Special labeling requirements	Guidelines

Source: U.S. FDA 2018a.

Example Device Classification: The Wheelchair

While these regulations and guidelines are helpful, looking at examples of classified devices often helps to illustrate them more clearly. Examples of Class I devices are listed in Table 7.1, where you can intuitively see that the degree of risk is low for use of a bedpan and perhaps moderate for the use of a wheelchair. Likewise, you can imagine the higher level of risk associated with a needle or a surgical stapler. Interestingly, the wheelchair is an example of a device that is a Class I device if it is not powered but a Class II device if powered. In this example, because the wheelchair is powered, special controls are needed to provide reasonable assurances of safety and effectiveness. General controls are not enough, so additional information is needed to provide those assurances. None of us wants to see Granny in her wheelchair drag racing with an ambulance because it is overpowered or crashing because the brakes aren't strong enough to stop it! Thus special controls may include performance standards for the motor and the brakes.

Although this example is facetious, it is not far from the truth. In 2008, a powered wheelchair was recalled because its stopping distance was longer than expected and could cause injury to the user (U.S. FDA 2020d). Because there were special controls regarding the performance standards of the brakes, the device was recalled, and the company corrected the issue, thereby ensuring the safety of users.

Labeling Example: The Wheelchair

Earlier it was mentioned that the FDA aims to ensure access to accurate information with labeling requirements. In Chapter 3, we learned about the charlatans who made false claims about the efficacy of their drugs and potions. We also learned of some horrific consequences when people used drugs that did not list the harmful ingredients they contained on the label.

Regulations are far more stringent today, and hopefully, they help us to avoid the catastrophic outcomes of the past. Still, in your career, you will frequently encounter labeling issues that may seem minor but are still important. Continuing with the wheelchair example, let's look at a typical labeling issue that you might encounter today.

In 2012, there were reports of injury when the padded swing-away armrest was used for full body weight on certain models of the Sunrise Medical wheelchair. Even though the owner's manual stated that the swing-away arms were not intended for full body weight, users were not aware and were getting hurt when they leaned on them to get up and the arms gave way. This is not surprising; honestly, when was the last time you fully read the owner's manual to anything? Sunrise Medical took appropriate action by sending a safety notice letter to all affected customers.

The letter identified the affected product, problem, and actions to be taken. It requested that clinicians/suppliers post the safety notice in a conspicuous location, making customers and therapists aware of the issue. It also encouraged end users to read the owner's manual and follow the recommendations for proper use (U.S. FDA 2020d). With this action, the recall was terminated because reasonable assurances for the safety of users were restored.

Exempt

Some medical devices are **exempt**. Most Class I devices and several Class II devices meet this criterion. A device may be exempt because there are specific regulations of exemption for that device or it has been grandfathered. A grandfathered device is officially a **preamendment device**; it is one that has been legally marketed in the United States since before the Medical Device Amendments were enacted in 1976 (see Chapters 3 and 5). For a device to be grandfathered, it must not have been significantly changed or modified since then, and there can be no regulations published by the FDA after 1976 requiring the manufacturer of the device to apply for premarket approval.

If a device is exempt, you do not have to apply to the FDA for permission to market that device, but you still must register your organization and also list your product with the FDA. Regulations for good manufacturing practices (GMPs) and product labeling still apply. All devices classified as exempt are subject to the limitations on exemptions. Limitations on device exemptions are covered under 21 **Code of Federal Regulations (CFR)** Parts 862–892. A current listing of medical device exemptions may be found on the FDA website (U.S. FDA 2020a).

Determining Your Regulatory Strategy

If your device is not exempt, you will need to apply to the FDA for clearance or approval before you can market your device in the United States. There are primarily two different types of applications. The first and generally less rigorous pathway is the **premarket notification** (**PMN**). This is more commonly known as the **510(k)** process and is generally reserved for Class I or Class II devices. For Class III devices, where the risk level is high, such as a pacemaker that supports or sustains human life or breast implants that may present a potential unreasonable risk of illness or injury, general and special controls are insufficient to provide reasonable assurance of the safety and effectiveness, so a **premarket approval application** (**PMA**) will be required. The PMA process is designed to help build and evaluate evidence that provides those reasonable assurances as well as better understand the risks involved. One exception to this is if your device is a preamendments device (on the market prior to passage of the Medical Device Amendments in 1976 or substantially equivalent to such a device) and PMAs have not been called for. In that case, a 510(k) process will be the route to market.

Cost and Time to Market (PMA versus 510(k))

A simple comparison of the average cost and time to market for each pathway will give you an appreciation for the motivation by device manufacturers to obtain the lowest classification possible for their new devices. Table 7.4 lists the average cost and time-to-market results from a 2010 survey involving 200 medical device manufacturers.

TABLE 7.4   Comparison of the Average Cost and Time to Market between a 510(k) and PMA for 200 U.S. Medical Device Manufacturers

Average	510(k)	PMA
Cost from concept to clearance/approval	$31 million	$94 million
Time from first FDA communcation to approval	31 months	54 months
Cost per additional month if delayed	$520,000	$750,000

Source: Makower, Meer, and Denend 2010.

You can see from the table that the average cost to obtain a PMA is three times that for a 510(k) device and that the average time involved with the FDA is nearly doubled. Because these values are so huge, many manufacturers would find the PMA process to be cost prohibitive. There is also a greater deal of uncertainty in the PMA process, which is of great concern to a company or an investor, especially considering the huge monthly cost of delays.

Clearance versus Approval

In the 510(k) process, you, the **sponsor**, provide a premarket notification submission to the FDA. The FDA then reviews it and provides clearance to market the device. If your device was required to undergo the PMA process, then you, the sponsor, would have to submit a premarket approval application to the FDA. When the FDA approves the application, then the device is said to be approved (U.S. FDA 2020e).

It is the sponsor's responsibility to provide all required evidence. Therefore, it is critical that you and your team understand the successes and failures of similar devices and components on the market or in development so that you can provide evidence that addresses any issues or sensitivities that may exist related to your product. Ultimately, the FDA determines your device classification, and if you do not provide reasonable assurances of safety and effectiveness, the agency will continue to ask questions before granting clearance, and it could determine that your device is to be placed into a higher risk classification.

Determining Device Classification

The FDA website offers an abundance of information regarding device classification and has listed more than 1,700 different types of devices (www.fda.gov). For each of these devices, there is a section of the Code of Federal Regulations (CFR) that gives a general description including the intended use, the device class, and marketing requirements. These CFR descriptions can be found in 21 CFR 862–892, and each is associated with one of 16 medical specialty panels, such as orthopedic or neurologic. As a first step, you should review the panels to determine which CFR regulates your device. Your device should meet the definition contained in the associated CFR (U.S. FDA 2018a).

The next step would be to proceed to the guidance for market submission that is recommended based on the CFR and device classification, such as the premarket notification page for 510(k) devices or the premarket approval page for Class III devices (U.S. FDA 2020a).

Using a Device before It Can Be Legally Marketed

In special cases, a medical device may be used on certain patients before it has been cleared or approved for market. For example, if you want to collect clinical data on your device, you will have to perform a clinical research study. In this case, you will need to study your device on/in human subjects before it is approved. In order to do so, you must obtain an **investigational device exemption** (**IDE**). You must also obtain **institutional review board** (**IRB**) approval.

To do so, you would submit an IDE application to the FDA and also submit all the necessary documentation to your IRB. The clinical research procedure will be discussed in more detail in Chapters 8 and 9. What is important to note here, however, is that the device may be used prior to FDA approval only in the context of an IDE and only after the IDE is issued and IRB approval has been granted. Guidance for IDEs can be found on the FDA website (U.S. FDA 2019).

There is a special case for a **humanitarian use device** (**HUD**). This is "a medical device intended to benefit patients in the treatment or diagnosis of a disease or condition that affects or is manifested in not more than 8,000 individuals in the United States per year" (U.S. Department of Health and Human Services [DHHS] and FDA 2019). It is aimed at enabling the use of a medical device to help cure someone who has a rare condition for which it would be impossible to collect sufficient clinical evidence to obtain approval via a PMA. This is a special case that includes restrictions on use and profitability.

Nonsignificant Risk Device

There may be a case where you believe that your device poses minimal risk to human subjects in your investigation. Maybe your study involves testing a new blood pressure monitor, and your study involves simply putting a healthy subject on a treadmill for two minutes and taking blood pressure data. You might feel that there is minimal risk to just about anybody you would involve in this study and that you have been responsible to engage doctors and clinicians to properly evaluate and care for your subjects.

You, as the device sponsor, are responsible for making the initial risk determination. You might feel that your device poses very little risk to a human subject in a particular study. You may be able to streamline your process, skipping the need to obtain an IDE from the FDA. You would establish a claim that your device is of **nonsignificant risk** (**NSR**) to human subjects in the study that you have designed, and you would present your rationale for that risk determination to your IRB. It is then the IRB's job to review your NSR determination for the study that you propose. Your claim of NSR must be verified by your IRB. If the IRB agrees with you, the sponsor, no change is needed, and you may proceed to conduct the study upon IRB approval. If the IRB disagrees with you, it will require modification of your study, and depending on whether the modifications are approved or not, you then must apply for an IDE with the FDA. Note that if the FDA has already made the NSR determination for the study, the agency's determination is final.

Is My Device Class II or Class III?

When you develop a new product, you will often encounter the question, "Is this a 510(k) device?" Obviously, the 510(k) application, because it is significantly less rigorous, is preferred over a PMA. As discussed previously, it will take less time, effort, and cost to bring the product to market. But what happens if your device is novel and does not seem to fit into one of the established panels? Practically speaking, because special controls are device specific, it is often difficult to be certain what body of evidence will provide reasonable assurances of safety and effectiveness.

When planning a regulatory strategy for the introduction of a new device, you will focus a lot of effort on answering this question. If you can prepare evidence that effectively demonstrates that your device may be cleared though with a 510(k) process, you will save a lot of time and money along your regulatory pathway. If not, the longer and costlier PMA is required to market the product in the United States. Therefore, it is very important that you understand the 510(k) process.

510(k) Process

I'll bet that you've wondered why it is called a 510(k). In the FDCA of 1938 (see Chapter 3), there is a Section 510(k) that states that device manufacturers who are not exempt must register with the FDA and, at least 90 days in advance, notify FDA of their intent to market a medical device. This is known as premarket notification (PMN). This allows the FDA to determine whether the device is **substantially equivalent** to a device that is already classified and legally marketed in the United States—a **predicate** device. A PMN is required for a new device as well for the reintroduction of a device if significant changes or modifications have been made to it.

Thus, if you can demonstrate that your new device is substantially equivalent to a predicate or several predicate devices, the new device may be reviewed and cleared via the PMN, aka 510(k), pathways. You can search the FDA databases to see if there are similar devices to yours that have already been cleared for market with a 510(k). Both the guidelines for 510(k) clearances (U.S. FDA 2018) and the current regulations describing the procedures may be found on the FDA website (U.S. DHHS et al. 2019).

Substantial Equivalence

Substantial equivalence means that the new device is at least as safe and effective as the predicate. "Substantial equivalence is established with respect to intended use, design, energy used or delivered, materials, chemical composition, manufacturing process, performance, safety, effectiveness, labeling, biocompatibility, standards, and other characteristics, as applicable" (U.S. FDA 2020). The FDA definitions for substantial equivalence are listed in Table 7.5.

TABLE 7.5   The FDA's Definitions of Substantial Equivalence

"A device is substantially equivalent if, in comparison to a predicate, it:
• has the same intended use as the predicate; and
• has the same technological characteristics as the predicate;
or
• has the same intended use as the predicate; and
• had different technological characteristics and does not raise different questions of safety and effectiveness; and
• the information submitted to FDA demonstrates that the device is at least as safe and effective as the legally marketed device."

Source: U.S. FDA 2020b.

It is very important to recognize that a claim of substantial equivalence does not mean that the new and predicate devices must be identical. Remember that in striving for innovation, a new device should be better in some way. Let's look at electrical spinal cord stimulators as an example. The early models used a battery outside the body. Some newer models have batteries that are implanted in the body. If the manufacturer of a new model provides sufficient evidence regarding the safety of the implantable battery and shows that the rest of the technology is equivalent to the older model, that manufacturer may be eligible for a 510(k) process for the new model. This is not a perfect process, however. Sometimes there are grave consequences to the user if a critical aspect is overlooked. You will learn more in Chapter 13 about spinal cord stimulators and some of the problems associated with them.

Advanced (But Important to Know About) Strategies

De Novo

If you cannot find a predicate device but you feel that with special controls you can provide a reasonable assurance that the product is safe and effective for its intended use, you can submit an application via the **de novo** process. The guidelines for this process are less specific and usually include a lot of early informal dialog with the FDA, including a presubmission discussion. Generally, a device manufacturer would engage an expert regulatory consultant to guide it through such a process. However, as a medical device designer, you should be aware that this pathway exists.

Fast Tracking a Class III Device: 513(g)

If you are working at a startup company that has developed a new knee implant, you may believe that the implant performs better than all the products currently available on the market. How could you get your new product to market using the PMN pathway, even though this type of device has been classified as a high-risk Class III device? You know that many knee implants have been on the market for a long time and that most have been safe and effective, and you feel that your device has only minor differences from those already on the market. Well, you can petition the FDA for information and a recommendation as to whether you might be able to fast track your application instead of going through the whole PMA process. The FDA will respond to your petition with recommendations; these recommendations are nonbinding but generally very helpful.

This was the case in the late 1990s. Although many companies offered a good knee implant and the products were safe and performed well, there was a need to offer new, smaller sizes. Until that time, these knee implants were designed primarily for men. They were indicated for use in women, often quite successfully, but in some cases the implants were too large for women of smaller stature, and it was recognized that the implant would perform better—and be more accessible to a broader population—if it were available in smaller sizes.

In this case, even though the original knee implants were approved as Class III, the new, smaller implants could be cleared via a PMN subject to American Society for Testing and Materials testing and other special controls that had been established for that type of device. The procedure to obtain this type of information from the FDA is called a 513(g) process, and the agency provides guidance for this process (U.S. DHHS et al. 2017).

Multiple Predicates

In some cases, you may choose to use multiple predicates, as long as those predicates have the *same intended use* as the new device *and* satisfy the other criteria for substantial equivalence. This would be the case if you are combining features from two or more predicate devices with the same intended use into a single new device. If Company J and Company Z already have knee implants on the market and your new knee implant simply combines what you believe are the best features of each, you may consider both as predicates if their devices and yours all have the same intended use. You would still need to provide information that your new device is just as safe and effective as those predicate devices and especially discuss how the new combination of features poses no more risk than the current devices.

You might also choose multiple predicates if you are seeking to market your device with more than one intended use or if you are seeking more than one indication under the same intended use. But you may not use split predicates. For example, suppose that you are planning to market a new electrical stimulation device to manage pain and you identify two legally marketed electrical stimulation devices that are similar to yours. Device A delivers a different signal from your new device to manage pain (it is technologically different but has the same intended use), and Device B delivers the same electrical signal as yours does but the indication is for bone growth stimulation (it is technologically the same as your new device, but the indication for use is different). In this scenario, you could not split the requirements for a predicate among different devices such as Device A and Device B.

Determining predicates can be a very complex exercise. The FDA has prepared very helpful guidance documents that are posted in the FDA website. Be sure to visit the website (www.fda.gov) and review the guidance documents related to the process you wish to consider. Also, you may contact the FDA to obtain guidance prior to submission, but keep in mind, again, that none of the recommendations are binding.

Most companies will have experts on staff who specialize in these regulatory procedures, or they will hire consultants with expertise to guide them through this process. In order to effectively recognize and communicate the critical elements of your design project that may affect the regulatory process, all team members should have an appreciation for the process requirements and the rationale for such.

Outside the United States

One commonly practiced tactic is to obtain clinical evidence **outside the United States (OUS)**. Depending on the site and economics of that location, a clinical study conducted OUS can cost 30 to 50 percent less than that of a comparable US study (Hogan 2015). Studies conducted OUS still must maintain procedural and data integrity and must also protect human subjects in accordance with good clinical practices (GCPs). Still, given the hundreds of thousands to millions of dollars that can be saved, it is no wonder that between approximately 30 and 45 percent of FDA applications included foreign clinical data in the period of 2012–2015 (Hogan 2015). This trend is expected to continue, and in recognition of this practice, the FDA prepared guidelines that can be found on its website (U.S. FDA 2020b).

CE Mark

Some manufacturers even introduce a product in other countries before marketing the product in the United States. This is done most commonly in Europe, which has a very good track record of ensuring overall safety and effectiveness of devices under its authority and where there are many similarities to the FDA processes. Though not identical to FDA clearance or approval, a **CE mark** is required before products may be marketed in the European Economic Area. CE stands for Conformité Européenne, and it is the manufacturer's declaration that it has conformed with health, safety, and environmental protection standards for that product.

General efficiency of the European agency is cited as the rationale for this strategy. Comparing the relative timeframe from first agency communication to clearance/approval/certification, you can see that it takes about four to five times as long to obtain permission to market in the United States compared with Europe (Table 7.6).

TABLE 7.6   Comparison of the Time to Market between the United States and the European Union for the Same Device

U.S. companies reported average time from first FDA communication to market clearance/approval	**FDA 510(k)**	**CE**	**FDA PMA**	**CE**
	31 months	7 months	54 months	11 months

Source: Makower, Meer, and Denend 2010.

If we look at the more than $500,000 per month that U.S. companies claim, on average, that it costs them to obtain a 510(k) and consider the average two additional years it takes to get 510(k) clearance after the CE mark, we see how the CE mark would save those companies $12 million! Of course, there could be added costs and challenges when dealing with travel and other cultures, but for certain types of products, launching in Europe can significantly reduce cost and time to market.

Summary

The FDA may ask for clinical evidence in either a 510(k) or a PMA. Usually, substantially more evidence is required for a PMA. The cost for a clinical trial can range between $1 million and $10 million and sometimes even more (Hogan 2015). If you consider the time required to develop a clinical trial, enroll subjects, and collect and evaluate data, you can begin to see why the timeframe becomes so much longer for a PMA. This is the biggest reason why, on average, a PMA takes so much longer and costs so much more.

Consequences of a Costly FDA Review Process

More than 80 percent of medical devices cleared or approved by the FDA in 2017 were cleared through a 510(k) process. The high percentage of 510(k) approvals compared with PMAs has been similar in several prior years as well. You should wonder what this means for the FDA's mission to encourage special innovations. Clearly, it is a challenging tradeoff to have reasonable assurances of safety and effectiveness but also to keep costs from being prohibitive. Some people argue that we may be safe at the expense of squelching or delaying the introduction of new ideas; others argue that we are not safe enough. Over the years, there have been many examples that support both sides of this argument. Keep in mind that the stakes in this tradeoff can be very high; recall what happened when thalidomide was introduced in parts of the world without having good clinical evidence of its safety (see Chapter 4). Even with all we know today, risks can be misjudged. Some current examples will be discussed in later chapters.

Study Questions

1. Outline the three classes of medical devices and give an example of each not given in the chapter.
2. What is labeling as it pertains to medical devices? Why is it important?
3. What is substantial equivalence?
4. What is a 510(k)? Why is it a desirable classification or not?
5. Why do some medical device companies try to sell their products in Europe before they launch them in the United States?

Thought Questions

1. What techniques can be used to streamline the use and regulatory approval process of a Class III medical device?
2. Define PMA. PMAs have both value and risk. Why is this?
3. What is the difference between FDA clearance and FDA approval?
4. Explain the concept of medical predicates in a way someone outside the field could understand.
5. How does the value proposition for a product change between the different device classifications?

References

Hogan, Janice. 2015. "How Will Conducting a Medical Device Clinical Trial Outside the U.S. Impact Your FDA Approval?" *Medical Device Online*, June 8, 2015. www.meddeviceonline .com/doc/how-will-conducting-a-medical-device-clinical-trial-outside-the-u-s-impact-your -fda-approval-0001.

Makower, Josh, Aabed Meer, and Lyn Denend. 2010. "FDA Impact on U.S. Medical Technology Innovation." www.advamed.org/sites/default/files/resource/30_10_11_10_2010_ Study_CAgenda_makowerreportfinal.pdf.

U.S. Department of Health and Human Services (DHHS), Food and Drug Administration (FDA), Center for Devices and Radiological Health, and Center for Biologics Evaluation and Research. 2014. "The 510(k) Program: Evaluating Substantial Equivalence in Premarket Notifications." Washington, DC, Juky 28. www.fda.gov/media/82395/download.

———. 2017. "De Novo Classification Process (Evaluation of Automatic Class III Designation): Guidance for Industry and Food and Drug Administration Staff." Washington, DC, October 30. www.fda.gov/downloads/MedicalDevices/DeviceRegulationandGuidance/ GuidanceDocuments/ucm080197.pdf.

———. 2019. "Humanitarian Device Exemption (HDE) Program: Guidance for Industry and Food and Drug Administration Staff." www.fda.gov/media/74307/download.

U.S. Food and Drug Administration (FDA). 2018a. "Device Classification Panels." Washington, DC. www.fda.gov/medical-devices/classify-your-medical-device/device-classification-panels.

———. 2018b. "Regulatory Controls." Washington, DC, March 27. www.fda.gov/medical -devices/overview-device-regulation/regulatory-controls.

———. 2018c. "510(k) Clearances." Washington, DC, September 4. www.fda.gov/medical -devices/device-approvals-denials-and-clearances/510k-clearances.

———. 2020a. CFR: Code of Federal Regulations Title 21: Medical devices. Washington, DC, April 1. www.accessdata.fda.gov/scripts/cdrh/cfdocs/cfcfr/CFRSearch.cfm?fr=807.92.

———. 2020b. "Consumers (Medical Devices)." Washington, DC. www.fda.gov/medical -devices/resources-you-medical-devices/consumers-medical-devices.

———. 2020c. "IDE Application." Washington, DC, November 25. www.fda.gov/medical -devices/investigational-device-exemption-ide/ide-application.

———. 2020d. "Medical Device Recalls." Washington, DC, November 20. www.accessdata .fda.gov/scripts/cdrh/cfdocs/cfRES/res.cfm.

————. 2020e. "Premarket Notification 510(K)." Washington, DC, March 13. www.fda.gov/medical-devices/premarket-submissions/premarket-notification-510k.

————. 2021a. "Medical Device Exemptions 510(k) and GMP Requirements." Washington, DC, January 1. www.accessdata.fda.gov/scripts/cdrh/cfdocs/cfpcd/315.cfm.

————. 2021b. "Recent Final Medical Device Guidance Documents." Washington, DC, January 14. www.fda.gov/medical-devices/guidance-documents-medical-devices-and-radiation-emitting-products/recent-final-medical-device-guidance-documents.

CHAPTER 8

Foundations of Clinical and Preclinical Research

Learning Objectives

Gain a foundation in the rules and procedures of clinical research.

Appreciate the risk and damage that can be caused by improperly conducted clinical research.

Learn about the humane use of animals in research.

Learn about the selection of animal type for preclinical research.

Understand the phases of clinical research.

New Terms

preclinical research

clinical research

Collaborative Institutional Training Initiative (CITI)

American Association for Laboratory Animal Science (AALAS)

institutional review board (IRB)

Office for Human Research Protections (OHRP)

Department of Health and Human Services (DHHS)

informed consent

Institutional Animal Care and Use Committee (IACUC)

Animal Welfare Act

Public Health Service Policy on Humane Care and Use of Laboratory Animals

critical-sized defects

phase I, II, III, and IV clinical trials

comorbidity

statistically significant

pivotal study

Living Subjects Research

If you are designing medical devices, it is almost inevitable that you will need to perform tests on live subjects. Whether those subjects are animals or humans, you must adhere to the highest ethical standards. You must satisfy the requirements of all the governing bodies that oversee **preclinical research** for animals and **clinical research** for humans.

As a device designer, you must have a good understanding of the rules and procedures in clinical research. Even if you are not directly involved with the research subjects, you must know and have an appreciation for the constraints and limitations of live subject testing and be able to interpret the results obtained from the tests.

If you have not had experience with research on live subjects, this may be daunting. Conversely, without training, it is easy to miss small details that could have costly, immoral, or deadly consequences. It takes many years of experience and training to become an effective clinical researcher. *No matter how harmless or simple a test might seem, research should never be conducted on live subjects without complete authorization, nor should it be performed by anyone without appropriate credentials.*

Companies and research institutions that conduct clinical research have specialized personnel who have the necessary training to properly implement this kind of research. Sometimes they may hire consultants who are specially trained in the area of interest under study. Your organization will also have a local governing body that serves as the authority for research on human or animal subjects.

Many device design engineers do choose to participate directly in the research. If you wish to do this, you can obtain training from qualified clinical researchers in your organization. A good resource for clinical research training is the **Collaborative Institutional Training Initiative** (**CITI**: www.citiprogram.org). A good place to learn more about preclinical research is on the website for the **American Association for Laboratory Animal Science** (**AALAS**: www.aalas.org).

Foundation for Current Standards

People have been performing research on humans and animals since the dawn of time. There are many examples of meaningful and humane research throughout history. Unfortunately, however, many atrocities and much senseless destruction have been committed under the pretense of research. Harm has also been caused unintentionally by negligence and oversights in testing and trials. Recall from Chapter 4 that U.S. doctors were allowed to dispense a trial sample of thalidomide to patients who were pregnant, leading to children in the United

States being born with birth defects, despite the fact that the drug was not approved for sale in the United States. The ethical standards that we uphold today are based largely on our experiences, both good and bad.

Belmont Report

Our current standards for research involving human subjects arise primarily from the Belmont Report of 1979. In July 1974, the National Research Act was signed into law, and the National Commission for the Protection of Human Subjects of Biomedical and Behavioral Research was created. The commission was charged with identifying the basic ethical principles that should underlie the conduct of biomedical and behavioral research involving human subjects and developing guidelines that should be followed to ensure that such research is conducted in accordance with those principles. In 1979, the commission issued the Belmont Report in an attempt to summarize the basic ethical principles it identified (National Commission for the Protection of Human Subjects of Biomedical and Behavioural Research 1979).

Institutional Review Boards and the Office for Human Research Protections

Facilities that perform research on human subjects must formally designate a group to review and monitor biomedical research in accordance with Food and Drug Administration (FDA) regulations. That group is called an **institutional review board** (**IRB**) and must be appropriately constituted. The purpose of an IRB is to protect the rights and welfare of human subjects participating in research activities being conducted under its authority. All IRBs in the United States must register with the **Office for Human Research Protections** (**OHRP**) in the **Department of Health and Human Services** (**DHHS**).

Almost all research involving human subjects requires IRB approval. Even something as simple as a survey must be reviewed by your organization's IRB, except in certain circumstances. It is a good idea to contact your IRB anytime your research involves human subjects. Let the IRB make the determination as to whether its review and approval are required.

Role of the IRB

The IRB is responsible for reviewing research proposals before a project begins. During that review, the IRB determines whether the research project follows the ethical principles and federal regulations for the protection of human subjects. The IRB has the authority to approve, disapprove, or require modifications to a project.

Your organization's IRB is there to protect you and anyone in your organization who may be a human subject. If a professor in your school was performing research comparing the fitness levels of students, that professor would need to review the entire study plan with the IRB before he or she could ask you to participate in the study. It might be hard to believe that with all the fitness technology that's out there today, IRB approval is still needed to perform this type of study, but it must be proven that any research involving human subjects is ethical and that the subjects' interests are protected in accordance with federal regulations.

Adequate Protection

Let's say that the professor wanted to measure your heart rate and blood pressure before and after exercising on a treadmill. The risks and consequences must be thoroughly reviewed. Certainly, the professor would have to present a plan to make sure that you are healthy enough to participate in the research and that the exercise protocol is safe. He or she must make sure that a qualified doctor has cleared you for participation and that emergency resources and personnel are available in case you get hurt or don't feel well during the study. But the professor would also need to demonstrate that you will not be compared with others or be made to feel bad in any way about your performance results. A fit young athlete might not see the concern here, but imagine if the student was overweight or even just clumsy; you wouldn't want that student to feel inadequate. If any comparisons are to be made, they would have to be in a way that the subjects are deidentified and the individuals' confidentiality is maintained.

The IRB would also want to make sure that your religious values were not compromised. If the professor required all female subjects to remove their headwear, for example, this could be a problem for women in certain cultures. It is just as important that your subjects' social, emotional, and psychological welfare is protected, in addition to their physical well-being.

Also, it would not be appropriate for the professor to influence a person's decision to participate in the study. For example, if the professor threatened to fail you in his or her course if you choose not to participate or offered to give you an instant A if you volunteered, you would be vulnerable. Vulnerable subjects are not permitted. The IRB has special considerations for protected groups such as children, pregnant women, and prisoners and subjects with disabilities or who are disadvantaged.

These are just a few simple examples to describe the types of concerns the IRB could have. Basically, the IRB weighs the risks and benefits of the research. There could be physical, psychological, social, or economic risks. There could be benefits for the individual research subject or to society overall or to both. The IRB reviews the research from varied perspectives, including the perspective of the population being studied. How would the study population view the risks and benefits? Is any additional protection needed? You can learn more on the website for the OHRP (www.hhs.gov).

Constituency and Performance of the IRB

The DHHS has established requirements for IRB membership. Because the IRB must look out for your overall welfare, these recommendations are aimed at providing a board with varied backgrounds and perspectives. Collectively, these members must be sufficiently experienced and have expertise in the proposed area of research so that they can safeguard the rights and welfare of human subjects. The IRB should also be diverse in culture, gender, and race and be sensitive to community issues and attitudes. It also must be able to navigate laws and institutional commitments with regard to the research project.

An IRB should consist of at least five members. The membership should include one member who is scientific and at least one member whose primary interests are nonscientific. There must also be one member who is not affiliated with the institute in any other regard to serve as a community representative. An IRB may add a member with special expertise if required for a particular project (DHHS 1991). IRB members at your school might include a nurse, a psychologist, a humanities professor, a biomedical engineer researcher, a grant contract administrator, and a local community priest.

IRB Review Procedure

The professor who plans to do research must submit a complete description of the proposed project to the IRB. That description should include how subjects will be recruited and informed. A detailed **informed consent** form is required. All materials, equipment, tools, and research procedures must be described in detail. Using this information, the members of the IRB meet to discuss the risks and benefits of the research and whether the subjects are adequately protected.

Once the IRB decides that the benefits of the research are justified and the risks are minimized, it can approve the research project. IRB members might ask questions and go through another cycle or two of review. In fact, this is quite common. Even if the IRB approves the project, your institute's officials can disapprove. However, if the IRB disapproves the project, your institute's officials may not override that decision.

IRBs usually convene periodically; for example, it is common for an IRB to meet once a month. Organizations that conduct a lot of research involving human subjects may have IRBs that meet more frequently. You should know how often your organization's IRB meets and its timing for decisions or other requirements when you plan a project. It is not unusual for complex or riskier projects to go through multiple review cycles, meaning that it could take several months before IRB approval.

Note that the IRB not only conducts an initial review but also performs a continuing review of active projects. This is usually done annually or when there is a change to the project, but it could be done more often at the IRB's discretion (Office of the Commissioner, Office of Clinical Policy and Programs, Office of Clinical Policy 1998).

Animal Research Strategies

Humane Use of Animals and the Institutional Animal Care and Use Committee

It is very likely that you would test on animals in a research study before testing on humans. This would be called a *preclinical study*. The objective of a preclinical study is to collect data to review and support the safety of the new treatment before it is introduced to humans.

Facilities that perform research on animals must appoint an **Institutional Animal Care and Use Committee (IACUC)**. This requirement was established in accordance with the **Animal Welfare Act** (U.S. Department of Agriculture 2017) and the **Public Health Service Policy on Humane Care and Use of Laboratory Animals** (U.S. DHHS and National Institutes of Health 2015). The U.S. government has established principles for the utilization and care of vertebrate animals used in testing, research and training that summarize the framework for animal research activities. There are many laws, regulations, and policies in addition to federal guidelines. Some regulations are at the state and institutional levels, and professional societies also promote standards of conduct. When developing animal research plans, be sure to include your local IACUC or animal welfare committee to ensure compliance with all applicable requirements (AALAS 2020).

Animal Selection

When choosing an animal type for a study, you must consider physiologic, pathologic, metabolic, behavioral, economic, and even sociopolitical factors. Different animals have different likenesses to humans. A study with small animals would be more cost effective than one with large animals, but there may be limitations in likeness to humans. An example of sociopolitical pressures can be seen in the reactions of U.S. citizens to doing experiments with dogs, which are considered treasured animals by many. Because of sociopolitical pressures in the United States, exceptional care must be taken in the consideration of studies with dogs or primates, and the justification for use of these animals must be high with no other reasonable alternative.

Small Animals

Let's first consider the use of rodents. Rodents are relatively easy to obtain, inexpensive to maintain, and small and manageable in size. They also have short periods from one generation to the next, and there is lots of published historical data. With all these advantages, you might consider rodents for any study. But you have to consider certain limitations this species brings. One such limitation is that certain wild species of rodents can survive conditions that would make humans sick. For example, rats are able to survive while carrying fleas infected with the bubonic plague. This plague killed more than 50 million people as it swept through three continents in the fourteenth century. If you were studying survivability of an infectious disease, there would be concern about using rodents. Just because the rodent doesn't get sick does not mean that humans would not get sick.

If you were considering a bone-healing study, you might choose to evaluate **critical-sized defects**. This is a common type of study to look at adjuvant healing technologies. A critical-sized defect is the smallest wound in a bone that cannot heal spontaneously during the lifetime of an animal. Because the bone and wound size are larger in a rabbit than in a rodent, it is easier to be more surgically precise with the rabbit model. Also, relative to rodents, the bone remodeling characteristics of a rabbit are somewhat closer to those of humans. For these reasons, a rabbit model might be chosen for such a study. Rabbits are far less robust than rodents, however, and they frighten and die easily. This makes the rabbit model less manageable and more expensive than the rodent model.

Large Animals

At some point, however, you may not be able to perform an effective test on small animals. If you wanted to test a bone implant, for example, you would have to scale your design down significantly to fit into a rabbit. Thus, while the rabbit model may be useful for preliminary evaluations, you may need to test your implant on larger animals. If your device is a spinal implant, you would want an animal model with a spine that has size and characteristics similar to that of humans. Typically, goat, sheep, or cow models are used in such cases. It is not uncommon to perform early studies with small animals, and once preliminary information yields positive results, then conduct later and more specified studies in large animals to address some of the elements that were limited in the small animal study.

If you are performing a study to evaluate wound healing in the skin instead of bone, you might instead use a pig model. Pig skin resembles human skin in many ways. Studies involving burns, for example, are often done with pig models. In fact, the sun protection factor (SPF) in your sunscreen was established with pig models. But to illustrate a limitation, think about whether your SPF 30 sunscreen is really 30 times better than wearing no sunscreen. How many times have you gotten a sunburn even when you were wearing SPF 30 sunscreen? If you have, this is because you sweat, and pigs don't. For SPF 30 sunscreen to be as effective as rated, you would have to either not sweat or reapply it often.

Regardless of the type of large animal you are considering, you can easily imagine the expense of a study involving them. In addition, you can appreciate the need for special facilities. Often such studies are performed at research institutions that specialize in managing large animals, and your organization would establish a contract with such an institution to perform the study. Ultimately, the model that you select depends on many factors. A good designer is familiar with the strengths and limitations of the various models. You need to know the feasibility, costs, and the pros and cons of how the performance of your device, drug, or other technology in that model may be similar to or different from its performance in humans.

Clinical Trials

Despite all the good data and information that you might obtain with laboratory and animal studies, it is almost inevitable that you will need to perform testing in human subjects. While the previous tests have helped you to review many elements of your design, including safety and dosages, there comes a point in time when you need to prepare evidence of safety and efficacy in humans. Generally, you would perform such tests in phases, with each phase having a different purpose.

It is important to note that one size does not fit all for clinical trials. The study should be designed to answer questions that can support the risk-benefit analysis for the particular product. The sample population and sample size also depend on these questions. To some degree, the practicality must also be considered; for example, it would make sense that a study evaluating a new artificial heart would have fewer subjects than one that looks at a low-dose electrical frequency for back pain.

Furthermore, as you will see, device testing and drug testing are different, so the phases of testing differ in their objectives. Whether you are designing a device or a drug, it is important to have some basic understanding of the traditional clinical trial phases for drug testing, including the general objectives, types of subjects, and some relative sample sizes.

Phase I

The objective of a **phase I** trial is to determine the safe dose of a drug and also to study potential side effects. This phase involves a small group of people (20–80). The subjects in a phase I trial are healthy.

If a drug company wants to evaluate the safety of a new 6% medicated topical skin cream for the treatment of eczema, it would recruit 40 healthy people to participate in a phase I study. The company would evaluate the subjects to ensure that their general health is good and that they do not have eczema or otherwise sensitive skin or allergies. The subjects would be instructed to apply the cream three times a day on a specified area for 10 days. The site of application would be monitored closely for any changes or reactions. The overall health of the subjects would also be monitored.

The premise for using healthy subjects in a phase I study is that if a healthy person were to have a negative reaction, the drug certainly would not be safe for a person with a disease or otherwise compromised health. In this example, if redness or a rash developed at the application site on many subjects, the dose frequency or potency would need to be evaluated and new study performed. If no changes were observed at the application site but several subjects reported feeling nauseous while participating in the study, the dosage would still require revision. The next study may evaluate a 3% formula and change the frequency to two times per day.

In contrast, if most or all of the healthy subjects had no reaction and remained healthy throughout the study, this would provide good evidence that the prescribed dosage is reasonably safe. This does not mean that the dosage is safe for everyone. People with sensitive skin, allergies, or eczema may still have a negative reaction. This would need to be evaluated in the next phases.

Phase II

The objective of a **phase II** trial is to perform an initial evaluation of drug efficacy. The group of subjects is larger than in phase I, but it is still relatively small (100–300). The subjects in a phase II trial must have the disease that the drug is aimed to treat.

Continuing with our example, your drug company would now recruit 200 people who have eczema to apply the new 3% medicated cream twice daily on the site of the eczema. Once again, the site of application would be monitored closely for reactions or changes, and it would also be monitored for healing. If healing occurred very rapidly in the subjects or in a small subset of subjects, this may provide information to reduce the treatment protocol to five or seven days instead of 10 days of applications. Protocol evaluations such as this are also part of a phase II study.

Often the subjects in a phase II study are carefully selected. Even though they have the disease of interest for the study, it is desirable to select subjects who are otherwise healthy. This would give the best possible conditions for a positive outcome. A patient who has other health issues might have a negative response to the drug because his or her body is weakened by his or her **comorbidity.** As in any scientific study, it is important to try to have only one variable at this phase of evaluation.

Phase III

The objective of a **phase III** trial is to evaluate safety and efficacy of the drug in a representative population. The group of subjects is large, often about 1,000–3,000 people. It is highly desirable that this population be a **statistically significant** group. As in a phase II trial, the subjects in a phase III trial must have the disease that the drug is aimed to treat. The group should be fully representative of that population. For example, if you intend to treat men and women ages 18 to 64 years, the study population should represent these groups. If the drug is aimed at also treating the elderly, the study population would also need to include men and women older than age 64. The phase III study would collect information about efficacy as in the phase II study and would also compare the new drug to current standard treatments. In addition, more information about safety and side effects is collected.

The results of the phase III study are prepared within the premarket approval package that is submitted to the FDA. If approved, the company may begin to sell the product in the United States. This does not, however, complete the phases of clinical study. There is another phase after the drug is on the market.

Phase IV

A **phase IV** study is referred to as *postmarket surveillance* (PMS). In these studies, more information about the drug or treatment is collected, including a review of optimal use and further evaluation of side effects. Not all companies perform phase IV studies in a formal manner, but they certainly monitor the performance of their drug or treatment. Sometimes when the FDA approves a drug, it includes a recommendation that phase IV studies be conducted. There have been cases where a phase IV study has revealed new information about the safety of a drug or treatment. An example of such a case is discussed in Chapter 12.

While the FDA may approve a device and recommend that phase IV studies be performed, the onus of performing these studies is on the device manufacturer. You can imagine that once a company receives FDA approval, its efforts will focus on marketing and generating revenue. The incentive to perform additional research on a device that is already approved diminishes compared with that prior to approval. The FDA has limited resources to enforce the execution of a phase IV study, and often these are not developed with rigor or in a timely manner. In practice, you will find that a company will be more likely to execute a phase IV study if its believes that it will generate information to help increase market penetration, such as to gain further adoption or perhaps to expand indications.

Differences between Phases of Medical Device and Drug Clinical Trials

While medical device testing and pharmaceutical testing are very similar, there are some differences because of the nature of medical devices. For example, recall that phase I studies look at a healthy population. However, if you were developing a humeral rod, you would not break a healthy person's arm to test it. Thus, typically medical devices are not tested in healthy subjects. Instead, the first human studies are pilot studies that are designed to evaluate safety and performance of the device. They involve a small sample of the population with the condition or disease under study, say in the range of 10 to 30 subjects.

The next step in medical device testing is a large study aimed at determining effectiveness and to evaluate adverse events. This study usually involves a large population, and it is the "big" **pivotal study** from which you would collect data to support your FDA application. It is similar to a phase III drug study in this regard.

Typically, you would aim to include enough subjects to yield statistically significant data. Depending on the type and nature of the device, it may not be feasible to enroll enough subjects to reach that value. For example, while it would be reasonable to find several hundred subjects for a hip implant study, it could be extremely challenging to find even 30 subjects to study a new artificial heart. Therefore, the number of subjects must be reasonable. Typical enrollment may fall in the range of 30 to 500 subjects, but this is highly subjective, and you would usually consult experts and perhaps even the FDA to discuss the necessary population size for such a study.

Summary

The protection of living subjects during clinical research is paramount. The procedures that have been established must be followed. If the procedures are not followed, there may be grave consequences to a subject's well-being in terms of health, safety, and ethical or moral treatment. Any individual or company that violates the regulatory policies may be penalized.

To minimize risk of harm to living subjects, clinical studies on living subjects are often conducted in stages or phases. Each phase should be designed to yield specific information to determine whether further development is warranted. By evaluating the results in stages, investment costs can be minimized. If, for example, a drug makes a few healthy subjects sick during a phase 1 study, the drug can be dropped or reformulated before undergoing a more expensive study with a larger population.

The procedures for developing clinical studies of medical devices are similar to those for drugs with some modifications for practicality. For example, while a drug might be tested on a healthy subject, you couldn't replace the knee of a healthy human with a prosthetic. Generally, though, clinical data for a medical device would need to provide evidence similar to that which is comparable to a successful phase III drug trial in order to obtain PMA.

Study Questions

1. What is the difference between clinical and preclinical research?
2. Describe the Belmont Report and discuss its significance.
3. What is the role of an IRB? When does an IRB review projects?
4. What does IACUC stand for? What is its role?

Thought Questions

1. Compare the pros and cons of small versus large animal research. Would you ever use both types on the same project? When and why?
2. What requirements need to be met for an IRB board? Write an example list of who could be on an IRB board different from that outlined in the text.
3. What are the phases of drug-based clinical trials? How do they differ from medical device clinical trials?
4. Research the Tuskegee syphilis experiment. How did this experiment change the way that clinical trials are run in the United States? What charge is put on researchers who work with living populations?

References

American Association for Laboratory Animal Sciences. 2020. "U.S Government." Memphis, TN. www.aalas.org/iacuc/laws-policies-guidelines/us-government.

National Commission for the Protection of Human Subjects of Biomedical and Behavioural Research. 1979. "The Belmont Report." Washington, DC, April 18. https://doi.org/10.1021/bi00780a005.

Office of the Commissioner, Office of Clinical Policy and Programs, Office of Clinical Policy, Office of Good Clinical Practice. 1998. "Institutional Review Boards Frequently Asked Questions." U.S. Food and Drug Administration (FDA), Washington, DC. www.fda.gov/regulatory-information/search-fda-guidance-documents/institutional-review-boards-frequently-asked-questions.

U.S. Department of Agriculture. 2017. "Animal Welfare Act." Washington, DC. www.nal.usda.gov/awic/animal-welfare-act#:~:text=The Animal Welfare Act (AWA,or exhibited to the public.

U.S. Department of Health and Human Services (DHHS). 1991. "Federal Policy for the Protection of Human Subjects: Notice and Proposed Rules." *Federal Register*, Vol. 56.

———— and National Institutes of Health. 2015. "Public Health Service Policy on Humane Care and Use of Laboratory Animals." Office of Laboratory Animal Welfare, Washington, DC, pp. 1–22. http://grants.nih.gov/grants/olaw/references/PHSPolicyLabAnimals.pdf.

CHAPTER 9

Clinical Study Strategies

Learning Objectives

Learn about the time, cost, and resources that must be considered to effectively plan a clinical study.

Appreciate some of the critical factors in planning human subject enrollment, including patient privacy.

Identify multiple goals in performing clinical studies.

Appreciate the multidisciplinary collaboration needed to develop an effective clinical study.

New Terms

assent

Health Insurance Portability and Accountability Act (HIPAA) of 1996

Privacy Rule

knockout (KO)

p-value

biostatistician

inclusion and exclusion criteria

clinical adoption

key opinion leader (KOL)

posterolateral fusions

Impact of Clinical Study on Design Planning

If you have not yet developed a clinical study, it could be easy to underestimate the time, cost, and effort involved in completing a successful one. The pivotal clinical study could be the most expensive and time-consuming part of an overall product launch. Every detail needs to be planned as carefully as possible in order to ensure success. Imagine having spent millions of dollars on a study that took two years to complete, only to find out at the end that you didn't collect enough data or the right data needed to demonstrate effectiveness. This would surely prevent the product from reaching the market. In cases like this, most companies do not have the resources—or choose not to add resources—to continue with a project. Often a product that was not successful in its pivotal trial is abandoned.

Meanwhile, as the device designer, you need to ensure that the elements of a clinical study are planned well in order to ensure the overall success of your project. In particular, you need to be sure that you have planned for an appropriate amount of time and budget to support the clinical study and that the study collects all the data elements needed for you to show strong evidence of safety and effectiveness.

Also, you need to avoid the need to make any changes after the study has begun. Recall that the institutional review board (IRB) reviews all documents and plans related to the study. If there are any changes to anything related to the study, it must undergo IRB review. This means that you may not proceed with any changes until they are approved. This can contribute to significant delays for the study. If the changes involve an adjustment to the data being collected, the size of the cohort to which that data relate could be affected, and this could jeopardize the strength of your evidence. It often takes a significant amount of time to review and be certain about every detail of a study, but when you are planning a clinical study, it is critical that you allow enough time to review every detail so that changes are averted or, at worst, very minor.

As discussed in Chapter 8, many protections are in place to keep human subjects safe. *Human protections are serious and may not be circumvented under any conditions.* Sometimes consideration of these protections can add complexity to subject enrollment.

Time and Effort to Execute a Clinical Study

Let's continue with the example from Chapter 8 of the new humeral rod that you are developing. Your humeral rod study requires a surgical implantation and then three follow-up visits over the course of one year after surgery. At each visit, your protocol requires that the subject complete a questionnaire and undergo a series of evaluations lasting a few hours, including x-rays and measurements by a physical therapist. You can coordinate the first two follow-up visits with the subject's postoperative follow-up visits. The last follow-up visit, the one that is one year after surgery, must be set up independently from the surgeon.

Your research hospital is quite active in trauma surgery, and last month, 30 patients who would have qualified for your new device were treated. There are many reasons why you will only capture a portion of those candidates. One of the three trauma surgeons is eager to contribute as an investigator, but the other two have some reservations about trying new products. While the surgeon you have on board is willing to perform surgery with your device, he has stated that he will be busy prepping for surgery and will therefore be unable to speak to the patient before the surgery. Do not assume that it will take only a month to recruit 30 subjects for your pilot study.

One of the ethical principles outlined in the Belmont Report is that no one can be coerced or unduly induced to participate in your study. If you identify a candidate for your study, you would need to properly obtain informed consent from him or her. The general minimum requirements for informed consent are listed in Table 9.1. There may be cases where additional requirements are warranted. For example, you may have to identify the potential effects to an embryo if your study involves pregnant women. If your potential subject is a child or is otherwise unable to provide informed consent, you must obtain **assent** from the subject and then, in addition, obtain informed consent from the parent or legal representative.

Imagine that a candidate for your humeral rod has been hurt in a late-night car accident and has been brought to your hospital by ambulance. Will you have a qualified team member available to obtain informed consent from the candidate? If you do, is the patient in such a condition that she can lucidly receive informed consent instructions and give consent? Will the trauma surgeon be comfortable using your implant on this patient? If the answer to any of these questions is "No," then you would be unable to enroll this candidate.

Compensating Study Subjects

Subjects should not be unduly induced to participate. A small payment may be offered to subjects for their participation. The payment of subjects is a common practice, though ethical concerns remain regarding how a payment might motivate or impair judgement (Grady 2005). The value of that payment should be small enough to reduce or eliminate these ethical concerns. Therefore, it is customary to compensate subjects only for their time and inconvenience and perhaps to offset visit expenses such as travel and parking. Generally, the payment is only a couple hundred dollars or less for later-phase studies and maybe somewhat more for earlier-phase studies.

TABLE 9.1   General Requirements for Informed Consent According to the U.S. Department of Health and Human Services (§46.116)

Except as provided elsewhere in this policy, no investigator may involve a human being as a subject in research covered by this policy unless the investigator has obtained the legally effective informed consent of the subject or the subject's legally authorized representative. An investigator shall seek such consent only under circumstances that provide the prospective subject or the representative sufficient opportunity to consider whether or not to participate and that minimize the possibility of coercion or undue influence. The information that is given to the subject or the representative shall be in language understandable to the subject or the representative. No informed consent, whether oral or written, may include any exculpatory language through which the subject or the representative is made to waive or appear to waive any of the subject's legal rights, or releases or appears to release the investigator, the sponsor, the institution or its agents from liability for negligence.

(a) Basic elements of informed consent. Except as provided in paragraph (c) or (d) of this section, in seeking informed consent the following information shall be provided to each subject:

(1) A statement that the study involves research, an explanation of the purposes of the research and the expected duration of the subject's participation, a description of the procedures to be followed, and identification of any procedures which are experimental;

(2) A description of any reasonably foreseeable risks or discomforts to the subject;

(3) A description of any benefits to the subject or to others which may reasonably be expected from the research;

(4) A disclosure of appropriate alternative procedures or courses of treatment, if any, that might be advantageous to the subject;

(5) A statement describing the extent, if any, to which confidentiality of records identifying the subject will be maintained;

(6) For research involving more than minimal risk, an explanation as to whether any compensation and an explanation as to whether any medical treatments are available if injury occurs and, if so, what they consist of, or where further information may be obtained;

(7) An explanation of whom to contact for answers to pertinent questions about the research and research subjects' rights, and whom to contact in the event of a research-related injury to the subject; and

(8) A statement that participation is voluntary, refusal to participate will involve no penalty or loss of benefits to which the subject is otherwise entitled, and the subject may discontinue participation at any time without penalty or loss of benefits to which the subject is otherwise entitled.

Source: U.S. Department of Health and Human Services (DHHS) 1991.

Candidates are free to choose to participate in a clinical study. However, they also have the right to withdraw from the study at any time. You may have enrolled certain patients who find that they are uncomfortable with the physical therapy evaluation after surgery and choose to withdraw from the study. Or you may find it challenging to bring a subject back after a year for the final follow-up because that person is healthy and active and has no reason to come back to the hospital. Or another candidate may have a demanding job and small children and cannot get away from work to come into the hospital for several hours. In longer studies, subjects move or change their phone numbers and neglect to update their contact information.

These are all practical reasons that illustrate why you should not plan to complete recruitment of all 30 subjects in a month. In this example, you might allow six months to complete enrollment of 30 subjects. You should also put a recruitment team and schedule in place. It would make sense to coordinate the team's schedule with that of the trauma surgeon who is eager to participate. If there are periods in the day or week when your hospital receives a high incidence of humerus fracture from trauma, such as late Saturday night car accidents, it would be appropriate to aim for your team to be in place at those times.

Even though you have enrolled 30 subjects, you must be prepared to accept that some will not complete the study. If you really need data from 30 subjects, you might aim to enroll 40 or 50 subjects. Keep in mind that this would add to your recruitment period. If you know that you can enroll 30 subjects in six months (enrolling at a rate of five subjects per month), you would actually need 10 months to enroll 50 subjects.

Then, depending on your resources, it may take several weeks to complete the data analysis from your study. It is not unusual for this to take two to three months, or longer. From this example, you can also see how project costs add up. Institutes have differing ways that they compensate their staff (e.g., recruitment team, surgeon, physical therapist, etc.) for participation in a study. You should expect to pay compensation to the institute and any of the staff, especially if your plan requires them to provide care or services that are different from or beyond their normal course/scope of work.

Usually, a company seeking to develop and launch a new medical device will have a dedicated clinical study team, either in its ranks or as consultants. Still, the device designer must recognize the time, cost, and challenges involved in a clinical study in order to build and maintain an effective project plan. If you haven't learned all this, you might assume that it would only take a month or two to complete the study when, in fact, it would actually take a year just to get it started. Imagine how a 10-month delay would affect your project!

Patient Privacy and Health Insurance Portability and Accountability Act

Earlier it was stated that patient privacy must be respected. This is not simply a professional consideration. It is the law. When you perform a clinical study, the **Patient Privacy and Health Insurance Portability and Accountability Act (HIPPA) of 1996** mandates that you must follow procedures to prevent disclosure of any patient health information.

The protected information includes [i]ndividually identifiable health information, including demographic data that relates to:

- the individual's past, present, or future physical or mental health or condition,
- the provision of health care to the individual, or
- the past, present, or future payment for the provision of health care to the individual,
- and that identifies the individual or for which there is a reasonable basis to believe it can be used to identify the individual [U.S. Department of Health and Human Services (DHHS) 2013].

In other words, you cannot hide a subject's name but disclose an address or a neighborhood or any other information about the subject's location that could lead to identification of that individual. You cannot disclose information about someone's physical or mental health at any time, even if the condition was in the past or will be in the future. Similarly, you cannot disclose someone's insurance or other payment provisions.

When conducting clinical research, this type of information may only be disclosed to the study personnel who are listed in the study protocol that is approved by the IRB. The information must be kept safely locked at all times. The data must be effectively deidentified before they are analyzed or published. This usually involves assigning a number code to each subject. The documents that list the association between the subject and his or her respective code must also be kept securely locked.

In 1955, when President Eisenhower had his heart attack (see Chapter 2), no such laws were in place. While there had previously been a shroud of secrecy about a U.S. president's health condition, Eisenhower's staffers chose to change that executive policy in order to instill public trust and confidence. In the hours and days following the president's heart attack, the nation monitored his condition along with his doctors and caregivers. The public was readily informed of Eisenhower's diet, activity, and even his bowel movements (Lasby 1997). The staffers' strategy worked toward convincing the public of Eisenhower's honesty, but it did so at the expense of violating the president's right to privacy.

Perhaps President Eisenhower benefited from this approach, but not all cases end so well. If your company is downsizing and has to eliminate one of the two people who currently hold a senior engineer position, would it be fair for the company to make that decision based on the individual's health condition? If your grandfather's cardiologist and surgeon were riding an elevator together and your grandfather's nosey next-door neighbor was on that same elevator, would you want your grandfather's doctors discussing his condition on that elevator ride? Whether deliberate or inadvertent, improper disclosures such as these have occurred with detrimental consequences for individuals and their families.

Today we follow the **Privacy Rule** established by HIPAA. It is important to ensure that an individual's health information is protected but that it also flows to the caregiving team and providers effectively so that high-quality care can be delivered. Instilling national standards for the management of an individual's health information was the major goal of the Privacy Rule. In addition to requiring safeguards to protect the privacy of personal health information, it sets limits and conditions under which disclosure may be made without patient authorization. It also gives individuals the right to examine and obtain a copy of their health records and to request corrections (U.S. DHHS 2013).

Variations among a Population

Another significant challenge in clinical testing stems from the variations among a population. If you were testing a certain mechanical part, each part must be identical, and it should be a straightforward task to identify test articles, specific test methods/treatments, pass/fail criteria, and so on. Even with animal testing, subjects must be selected from a homogeneous species, and sometimes a species can be "designed" for a specific study. For example, test mice can be bred to eliminate certain genetic characteristics; they are known as **knockout** (**KO**) mice.

But we have none of these conveniences in clinical testing. It is not ethical or practical to "breed" humans to have uniform genetic characteristics. Instead, a study population must be selected carefully so that the performance of the device can be "assumed" to be similar among all members of that study population.

The human population varies greatly. In some ways, the variations are obvious. We vary by gender, height, weight, age, and race, to name a few. In most cases, you would consider that a young professional football linebacker would be treated differently from a petite older woman. A hip implant for linebacker Joe would have to be much larger and stronger than one for old Mrs. Jones. Also, young linebacker Joe needs a new hip that will last at least 40 years, whereas old Mrs. Jones needs one for no more than 20 years. Furthermore, surgical techniques for insertion of the device also may vary given that old Mrs. Jones' bone density may not be as good as linebacker Joe's.

Quickly, you can see that design and testing criteria vary greatly as you begin to account for the variability among the human population. Still, a medical device must be tested to prove that it is safe and effective in each of these patients as well as all the people along the spectrum between them. Therefore, study subjects must include several representatives; a sample must be selected for each group of patients that may respond differently to treatment.

To make matters trickier, even human subjects with very similar characteristics may respond differently to a treatment. Old Mrs. Jones might respond very well to her new hip replacement, whereas old Ms. Smith may experience chronic pain. Thus, one or two representative subjects are not enough to provide confidence that the new hip implant will provide good outcomes for all petite old women. Instead, the sample must contain enough subjects to offer confidence that the sample represents the entire group of petite old women. Likewise, another sample of subjects must represent patients like linebacker Joe. In addition, other sample groups must be selected to effectively represent other groups in the spectrum of patients for hip replacement.

Sample Size and Statistical Significance

Keeping in mind that clinical testing is costly and time consuming, most clinical research aims to enroll the smallest sample size possible but still large enough that it provides convincing evidence of safety and efficacy. Evidence can be convincing if it is statistically significant, and it is deemed statistically significant if the **p-value** is less than 0.05. Simply stated, if the p-value is less than the significance level (generally 0.05), there is reasonable evidence to suggest that there is a difference among the groups studied. This does not always mean that the difference is meaningful or relevant but simply that there is a very good probability that there is a difference.

To determine the size and composition of sample groups, you must look closely at each difference and also at the range of variation between the groups that are being compared. Usually, a **biostatistician** is engaged to perform the calculations for sample size determination. But the biostatistician cannot perform the calculations without working closely with the entire design team.

Before the trial, clinicians and engineers will need to provide clinical assumptions and performance expectations, including estimates of range variations. This must be done very carefully to ensure that the sample size will be big enough that the results are statistically significant when the calculations are done at the end of the study. If a variation is overlooked or the range is not estimated properly, the *p*-value could be altered and cause the results to be nonsignificant.

Inclusion and Exclusion Criteria

Sometimes it is necessary to identify certain characteristics or groups that do not qualify for a clinical study. Either the characteristic or group does not accurately represent the population to be studied or the device will not be appropriate for the specific condition for which a group requires treatment. For example, you would not want to ask a person with a bone infection to participate in your hip implant study because the infection could lead to complications that might reduce the success of the outcome, and this would not accurately reflect on the potential for success in a patient who does not have a bone infection. Similarly, you would not want to include a 90-year-old man in your study if your device is aimed at treating adults in their forties and fifties.

All **inclusion and exclusion criteria** must be clearly identified when you design a clinical study. Lack of clarity could lead to narrow indications for use, requests by regulatory reviewers for more data, or outright rejection of the study. Any of these conditions may negatively impact the marketability of your device.

Reasons for Clinical Testing

We have discussed the needs to perform clinical research to determine safety and efficacy. When done effectively, with a positive outcome, the clinical tests will provide some of the most important evidence needed to obtain regulatory approval. But just because your product has received regulatory approval for marketing does not mean that it is going to "fly off the shelves like hot cakes." Several other elements factor into the success of a product. Some of those can also be addressed within the context of clinical testing.

Clinical Adoption Key Opinion Leader

If the clinical community doesn't believe in your product, it will not support it, regardless of whether or not it has met all regulatory requirements. It is the clinical community's job to ensure that their patients are offered the best possible outcomes. While evidence of safety and efficacy go a long way toward supporting **clinical adoption**, that is not always enough to convince doctors or surgeons to switch to your product.

A typical concern might include long-term outcomes. Perhaps the doctor or surgeon has been using a certain brand of implant for 15 years and has had very good results; even their very first patients have no complaints. You would have to provide a compelling rationale as to why your device would be better than what that doctor or surgeon is using. Sometimes there are opportunities within the context of your clinical study to demonstrate how your product might offer improvements. For example, you could perform a phase IV study to demonstrate how much more range of motion patients with your humeral rod have compared with those who received the competitor's humeral rod.

Another typical example might be that your device is better for a certain subset or type of population. In the early 2000s, a spinal hardware sales representative was introducing a new spinal rod option for vertebral fusions. Instead of the traditional stainless steel systems, the rep was promoting an implant system made of titanium. The titanium implants were smaller and lighter for the same relative strength. They were also "immune" to magnetic resonance imaging. You might think it would make sense to use the titanium system universally, but whether the doctors or surgeons made their selections based on their own experiences or from an intuitive evaluation of the design, titanium systems were not quickly and universally adopted. At orthopedic conferences, there were reports of the titanium rods snapping in patients who underwent high impacts, such as having been in a car crash or sustaining a hard landing while skydiving. The patients who had steel rods might have had some bending, but the rods did not break, and this bending was far easier to manage. Thus, doctors or surgeons often stayed with their tried and true steel products despite the larger size and mass.

Still, one day that same surgical rep was helping a surgeon prepare for an upcoming scoliosis surgery on an eight-year-old boy. The surgery involved multiple levels of hardware, and the boy was very light and small. In this case, the surgeon and the sales rep discussed the device options, and the surgeon determined that the titanium system would be better for the young boy. This represents effective clinical adoption of the device for a subset of patients who are smaller.

While this was just one case, it could serve as a good example to build on. For instance, that surgeon may have another small patient next week or next month, and if the prior experience with the titanium implant was successful, the surgeon would likely choose it again for another patient with similar challenges. The surgeon also might share his or her success with colleagues in his or her practice and in his or her hospital.

If that same surgeon happened to be the most senior spine surgeon in the biggest hospital in a major city, he or she would likely also be considered a **key opinion leader** (**KOL**) among his or her peers. In this case, the surgeon's influence would be broad. He or she would commonly provide leadership as a speaker at medical conferences and as a clinical instructor, as well as being a decision maker in his or her institution. The surgeon would discuss his or her success with difficult surgical cases, and many younger or less experienced surgeons would then follow his or her lead.

As you design your clinical studies for your device, it is critical that you prepare evidence that will support a positive opinion among the clinicians who will use the device. In this way, the clinical adoption of your device would be highly dependent on the opinion of a KOL. You must keep in mind that the opinion of surgeons may be more broadly based than the Food and Drug Administration's requirements for evidence of safety and efficacy. For the surgeons, acceptance may have to do with ease of use, surgical time, access to the product, or several other practical factors. You should consider how you might prepare such evidence in the context of your clinical studies as well.

Reimbursement Evidence

If a medical device reaches the marketplace with demonstrated safety and efficacy, and with interest of adoption by clinicians, but it cannot be reimbursed by insurance, market penetration may be extremely limited. Such was the case with an artificial lumbar disk replacement. This is a cautionary tale illustrating the need to evaluate more than safety and efficacy during a clinical trial. Reimbursement elements must be tracked. There must be sufficient evidence for insurance providers to be confident that the intervention will improve health outcomes for patients.

Spinal Disk Arthroplasty

In 2004, the Food and Drug Administration (FDA) approved the first artificial spinal disk. The potential benefits to a patient were purported to be exceptional. Until the disk arthroplasty option was available, the only choice a patient had after a spinal fracture was to undergo a spinal fusion. Even without knowing much about spine surgery, you can appreciate the potential benefit of having a flexible disk in your back or neck compared with having a part of your spine fused.

In fact, several companies invested millions of dollars in the development of an artificial spinal disk, expecting blockbuster sales. In the early 2000s, there was a frantic race to market. The artificial spinal disk was the most anticipated orthopedic device of the twenty-first century. By 2003, DePuy had acquired the Charite disk for $325 million, Synthes had acquired the ProDisc for $175 million, and Medtronic had acquired the Bryan cervical disk for $270 million. The projected estimated market value for spinal disk arthroplasty was as much as $3 billion annually (Varona 2004, Awe et al. 2011).

Despite initial expectations, at its peak, disk arthroplasty procedures represented only 3 percent of **posterolateral fusions**. That number declined steadily over the next several years, decreasing to 2 percent by 2008 (Awe et al. 2011). While DePuy conducted a noninferiority study to demonstrate that the Charite disk was safe and effective compared with the BAK cage (lumbar fusion), it failed to convince insurers that it offers patients improved health outcomes.

To date, the Centers for Medicare and Medicaid Services (CMS) has found that lumbar artificial disk replacement (LADR) is not reasonable and necessary for the Medicare population over 60 years of age. Citing reasons such as narrow indications or patient selection criteria and limited data on long-term outcomes and complications, CMS is upholding a national noncoverage determination that was made in 2006 (Phurrough et al. 2006).

Typically, most private insurance companies in the United States follow CMS coverage decisions and thus do not provide coverage for LADR. This lack of reimbursement coverage is the most significant reason for the very limited success of this device.

Important Design Knowledge

Ultimately, you must provide convincing evidence that your device offers a value proposition. Very simply stated, insurers want to know that your device offers an economical solution with good chances for an improved outcome for the patients they insure. Therefore, in addition to preparing evidence of safety and efficacy, it makes sense to track costs, time, patient length of stay in the hospital, and other resources required to provide care and support using your device. In addition, short- and long-term outcomes, complications, and quality-of-life assessments should be monitored. The need to accurately track costs and expanded outcome measures can be overlooked, especially during earlier phases of clinical trials. Still, for a device to be successful in the marketplace, cost, resource, and outcome tracking is just as critical to the evidence pool as safety and efficacy data.

Summary

Clinical studies must be developed with care. The subjects' best interests must be considered at all times, and the recruitment, management, and retention of all participants can be challenging. In practical terms, a well-designed clinical study may require a significant amount of time and effort to implement. An outsider or a novice can easily overlook elements that can nullify any value from a clinical study. The strategies involved in planning and executing a clinical study that will meet the objectives of FDA scientists, KOLs, and reimbursement agencies may be complex and require detailed knowledge that spans several diverse disciplines. For this reason, it is highly recommended that you engage a clinical research specialist in the design of a clinical study.

Study Questions

1. What is the medical community's accepted *p*-value for significance?
2. Why is appealing to KOLs so important?
3. What was the main factor that lead to the Charite disk failing to have widespread market penetration?
4. Why is careful selection of study participants so important?

Thought Questions

1. Imagine that you are on a team that has designed a new type of portable ventilator. Outline a brief explanation of what your clinical trial would be looking for. What kinds of patients would you hope to recruit for your study?
2. How could current information technology practices conflict with HIPAA? How might information technology improve HIPAA?
3. Your organization has designed a new version of a shoulder joint for a prosthetic arm. Current "gold standard" devices last for a long time but limit motion to forward and backward movements only. Your joint has more degrees of freedom but a tradeoff in its relatively low durability. For what populations might you suggest that your joint could improve the quality of life when working with clinicians?

References

Awe, Olatilewa O., Mitchel G. Maltenfort, Srinivas Prasad, et al. 2011. "Impact of Total Disc Arthroplasty on the Surgical Management of Lumbar Degenerative Disc Disease: Analysis of the Nationwide Inpatient Sample from 2000 to 2008." *Surgical Neurology International* 2:139. https://doi.org/10.4103%2F2152-7806.85980.

Grady, Christine. 2005. "Payment of Clinical Research Subjects." *Journal of Clinical Investigation* 115 (7):1681–1687. https://doi.org/10.1172/JCI25694.example.

Lasby, Clarence G. 1997. *Eisenhower's Heart Attack: How Ike Beat Heart Disease and Held on to the Presidency.* Lawrence: University Press of Kansas.

Phurrough, Steve, Marcel E. Salive, Deirdre O'Connor, et al. 2006. "Decision Memo for Lumbar Artificial Disc Replacement." Centers for Medicare and Medicaid Services, Baltimore, MD.

U.S. Department of Health and Human Services (DHHS). 1991. "Federal Policy for the Protection of Human Subjects: Notice and Proposed Rules." *Federal Register*, Vol. 56. www.govinfo.gov/content/pkg/CFR-2016-title45-vol1/pdf/CFR-2016-title45-vol1-part46.pdf.

———. 2013. "Summary of the HIPAA privacy rule." Washington, DC. www.hhs.gov/hipaa/for-professionals/privacy/laws-regulations/index.html?language=en.

Varona, Michael. 2004. "Hot Fields, Hot Companies." *Medical Device and Diagnostic Industry*, January 1. https://pubmed.ncbi.nlm.nih.gov/22059134/.

CHAPTER 10

Kyphon and Reimbursement

Learning Objectives

Understand how/why medical insurance is structured in the United States.
Learn about how certain factors contribute to the differences in reimbursement rates.
Be able to recognize why private payers and Medicare are linked.
Learn about osteoporosis and treatment for vertebral compression fractures.

New Terms

kyphoplasty

vertebral compression fractures

outpatient

inpatient

osteoporosis

bone mineral density (BMD)

kyphotic deformity

conservative management

spinal decompression and fusion

percutaneous vertebroplasty

interventional radiologists

polymethylmethacrylate (PMMA)

extravasation

reimbursement

Medicare Economic Index (MEI)

Centers for Medicare and Medicaid Services (CMS)

relative value unit (RVU)

resource-based relative value scale (RBRVS)

American Medical Association (AMA)

geographic adjustment factor (GAF)

geographic practice cost index (GPCI)

The Lawsuit Involving Kyphon, Inc.

In July 2007, Medtronic, Inc., announced that it would acquire Kyphon, Inc., for $3.9 billion. Kyphon had established itself as a rapidly growing company specializing in **kyphoplasty**, a procedure for treating **vertebral compression fractures**. At the time of acquisition, the company had grown at a rate of 30 percent per year with profit margins as high as 90 percent (Sherman et al. 2008).

By May 2008, not even a year later, Medtronic settled a civil lawsuit on allegations of Medicare fraud (Williams Walsh 2008). The lawsuit claimed that Kyphon improperly persuaded hospitals to keep people overnight. Kyphoplasty is a minimally invasive procedure to repair small fractures of the spine and usually can be done in about an hour in an **outpatient** setting. However, if a patient was kept overnight, Medicare would reimburse hospitals more because the procedure was deemed to be more complex **inpatient** back surgery. While it is reasonable to consider that certain patients at high risk of complications should be treated in an inpatient setting, the suit included testimony from two whistle blowers, Charles M. Bates and Craig Patrick, who had worked for the company. They said that Kyphon had deliberately urged doctors to admit patients overnight, knowing that the admissions were unnecessary.

Osteoporosis

To fully appreciate the finesse in this case, you need to begin by understanding the disease condition that the kyphoplasty procedure was designed to treat. **Osteoporosis**, defined as low **bone mineral density** (**BMD**), is a major public health problem affecting older adults. Bones with lower density are more prone to fracture, even just from the wear and tear of daily life. It is estimated that about 1.5 million fractures that occur annually can be attributed to this condition, at a cost to the U.S. healthcare system of $5 billion to $10 billion annually. Nearly half of these fractures are vertebral fractures. The condition mostly affects the population segment of older women; 26 percent of women older than 50 years of age have a vertebral compression fracture, and by 80 years of age, the prevalence increases to 40 percent (McCall, Cole, andDailey 2008). The incidence and costs are certain to increase along with our aging population (Riggs and Melton 1995) the definition of osteoporosis should not require the presence of fractures but only a decrease in bone mass that is associated with an unacceptably high risk of fracture. In the USA, approximately 1.5 million fractures annually are attributable to osteoporosis: these include 700,000 vertebral fractures, 250,000 distal forearm (Colles).

Vertebral Compression Fractures

The leading cause of vertebral compression fractures is osteoporosis. A vertebral compression fracture can have multiple other issues associated with it, such as severe back pain and reduced vertebral height, which can lead to a **kyphotic deformity** (Figure 10.1). Until the mid-1980s, treatment methods included **conservative management**, which required a resting period of generally three weeks or longer and often had patients enduring severe back pain throughout the healing period. In other cases, **surgical decompression and fusion** were performed, but these surgeries required that hardware such as screws be implanted into the bone (Jha 2013). The surgery often failed in elderly patients because of exactly what had caused their fractures in the first place—their low BMD. The screws could not maintain their purchase in the sparsely filled, brittle bones.

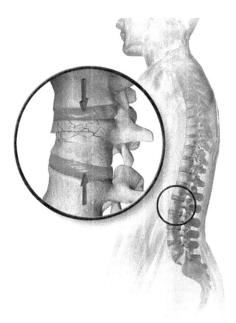

FIGURE 10.1   Kyphotic deformity results from compression fractures due to osteoporosis.

Vertebroplasty and Kyphoplasty

In 1984, the first **percutaneous vertebroplasty** was performed by a team of **interventional neuroradiologists** in France (Jha 2013). Percutaneous procedures are performed through the skin with only a small puncture created by a needle. The French physicians injected a bone cement made of **polymethylmethacrylate** (**PMMA**) into a patient's cervical spine and successfully alleviated his chronic pain. By 1993, physicians in the United States had begun using the technique. The procedure provided stability to the spine and became popular in part because it rapidly alleviated pain in about 80 percent of patients.

Despite the good efficacy of the vertebroplasty procedure, there were still some areas for improvement. The procedure was not effective in restoring the height that was lost during compression of the vertebral body. There is also some risk for bone cement to cause severe complications if it migrates to areas other than the site of treatment. This procedure required that the bone cement be injected at high pressure, which could cause **extravasation**. If the cement leaked into the patient's lungs or organs, it could result in embolization. It could even cause additional spinal injury if it leaked into undesired areas of the spine.

To address the need to restore the height of the vertebra, Mark Reiley, an orthopedic surgeon, introduced the idea of inserting an inflatable balloon in the vertebral body. The balloon is expanded to restore the vertebral body to its original height, and a cavity is created. The balloon is then removed, and bone cement is gradually injected into the cavity. The controlled injection of cement in this procedure presents less risk of extravasation. In approximately 20 minutes, the cement is dry, and the procedure is completed. The whole procedure lasts about 45 minutes. When it is completed, patients are able to walk immediately, and most patients experience pain relief as well as some restoration of vertebral height.

In 1994, Dr. Reiley and colleagues founded Kyphon, Inc., based on their kyphoplasty procedure. As mentioned earlier, the company enjoyed tremendous success for more than a decade. The company's aggressive marketing tactics enabled it to artificially drive up demand among hospitals. This led to the rapid growth and high profitability of the small startup company, making it a very attractive acquisition. It is worth noting that the lawsuit against Kyphon did not catch Medtronic by surprise; the company's due diligence concerning the acquisition was thorough. Medtronic still felt that Kyphon was a good strategic purchase. In the settlement, Medtronic agreed to pay the federal government $75 million plus interest and to enter into a corporate integrity agreement with the Office of Inspector General of the Department of Health and Human Services.

Profits and Abuse

You might wonder how such a situation was even possible. How could the identical kyphoplasty procedure be so much more expensive if the patient stays overnight in the hospital compared with if the patient went home after the procedure? Why would Kyphon representatives suggest that patients stay overnight in the hospital, even if most patients get up and walk immediately after the procedure is completed? After you read the rest of this chapter and Chapter 11, you will fully understand how and why this occurred.

But, as you already know, medicine is a big business. In this period, it was quite common for small companies to arise from a specific idea, just as Kyphon Inc. did. The exit strategy for many of these companies was to be acquired by a large company that was seeking a new product or idea to help it maintain a strong position in the market. In this case, Medtronic saw that the Kyphon product would support its leadership role in the spinal surgery space. Kyphon was an ideal target for acquisition, having not only a successful new product but also high profits and a rapid growth trajectory. Despite the $3.9 billion purchase fee, litigation, and a $75 million fine, it was still worthwhile for Medtronic to acquire Kyphon.

What Is Reimbursement?

As with any business, profits are paramount in the healthcare industry, so profits from medical products, procedures, and services ultimately drive a company's business. The payment for medical products and services is, however, unique to medical businesses. This is because most people have medical insurance and do not pay their medical bills directly. Instead, insurance providers pay a covered person's bill or the portion of the bill that has been agreed on between the provider and the licensed caregiver or institution. The caregiver or institute must account for every service or product provided to the patient using standardized codes and then submit its bill to the insurance company for **reimbursement**. Prices for each service or product are fixed in accordance with the standardized codes. This is a very highly structured and complex system that is constantly evolving. It might be easier to appreciate and understand this system if we look at how it developed.

Genesis of Modern Healthcare Reimbursement Concepts

Reimbursement structures for medical services have been implemented since ancient times. One of the earliest structured fee schedules is within the Code of Hammurabi (Table 10.1), which dates back to 1746 BC. Fees for surgical services were specified in the code, much like fees are specified today. In addition, the Code of Hammurabi required that diseases and therapies be documented. Furthermore, it prescribed legal action and penalties if expected outcomes were not met; our modern-day malpractice program resembles this aspect of the code.

Nearly 4,000 years after the Code of Hammurabi, up until the mid-twentieth century, the U.S. reimbursement system closely resembled those of ancient times. Fees were charged for services in accordance with a schedule of usual and customary charges. The insurance payer reimbursed the physician an agreed-on amount, and the patient paid any remaining balance.

TABLE 10.1   Hammurabi's Laws Dealing with Conditions Necessitating Surgical Activity

215. If a physician make a large incision with an operating knife and cure it, or if he open a tumor (over the eye) with an operating knife, and saves the eye, he shall receive ten shekels in money.

216. If the patient be a freed man, he receives five shekels.

217. If he be the slave of some one, his owner shall give the physician two shekels.

218. If a physician make a large incision with the operating knife, and kill him, or open a tumor with the operating knife, and cut out the eye, his hands shall be cut off.

219. If a physician make a large incision in the slave of a freed man, and kill him, he shall replace the slave with another slave.

220. If he had opened a tumor with the operating knife, and put out his eye, he shall pay half his value.

221. If a physician heals the broken bone or diseased soft part of a man, the patient shall pay the physician five shekels in money.

222. If he were a freed man he shall pay three shekels.

223. If he were a slave his owner shall pay the physician two shekels.

224. If a veterinary surgeon perform a serious operation on an ass or an ox, and cure it, the owner shall pay the surgeon one-sixth of a shekel as a fee.

225. If he perform a serious operation on an ass or an ox, and kill it, he shall pay the owner one-fourth of its value.

226. If a barber, without the knowledge of his master, cut the sign of a slave on a slave not to be sold, the hands of this barber shall be cut off.

227. If anyone deceive a barber, and have him mark a slave not for sale with the sign of a slave, he shall be put to death, and buried in his house. The barber shall swear: "I did not mark him wittingly," and shall be guiltless.

Source: Halwani and Takrouri 2006

Birth of Employer-Paid Health Insurance

Health insurance was not popular in the early twentieth century. Frankly, medicine was not very sophisticated, so there was not much expense involved. Most people paid a reasonable and manageable fee or traded in kind for a doctor's services.

In 1942, the United States found itself entrenched in World War II, and many workers were diverted to military service, creating a severe labor shortage. The United States was still recovering from the Great Depression of 1929. There was a fear that employers would compete for workers by continually raising salaries and that this would lead to devastating and uncontrolled inflation. An executive order was issued that created the Office of Economic Stabilization and froze wages (Carroll 2017).

Wages could not be used to attract workers at this time, so businesses found other ways. They began offering benefits for healthcare. By the mid-twentieth century, medicine had evolved to where some of the best technology and care could involve large expenses, making the notion of healthcare insurance attractive.

In 1943, the Internal Revenue Service made those benefits tax deductible. The result was that it was much cheaper to obtain healthcare insurance through an employer than independently. Precipitously, American employers found themselves in the business of providing health insurance despite having no background in healthcare.

Medicare Leads to Changes from the Customary Fee-for-Service Model

With the intention to ensure that the elderly, low-income children, and individuals with disabilities had access to healthcare services, President Lyndon Johnson signed into law Medicare and Medicaid as part of the Social Security Act in 1965. At that time, only half of U.S. citizens aged 65 and older had health insurance coverage, and this was the group that was most likely to be living in poverty (Centers for Medicare and Medicade Services [CMS] 2015). For the next decade, Medicare payment methodology was the same as others: usual and customary charges were reimbursed, and patients were billed for any remaining balance. By 1972, however, concerns arose about rising physician fees, and Congress mandated that any added fee be limited to something that is "reasonable" based on a calculation that accounts for any increased costs related to the physicians' operating expenses and earnings levels. This calculation became effective in 1975 and is updated every year; it is known as the **Medicare Economic Index** (**MEI**).

Medicare Sets Billing Guidelines

You might be wondering why the focus is on Medicare. After all, nearly half of all Americans have private healthcare insurance through their employers and less than 20 percent of the U.S. population is enrolled in Medicare (Figure 10.2). The fact is that today many private payers follow Medicare's reimbursement methods, so while Medicare decisions directly affect only a small part of the population, they impact almost all who are insured, which is nearly 300 million people! Today the United States spends more than $3.3 trillion annually on healthcare, and Medicare is run by a government agency, the **Centers for Medicare and Medicaid Services** (**CMS**: www.cms.gov). Therefore, prices for medical services and products are governed primarily by Congress via the payment methodology that has been built for Medicare.

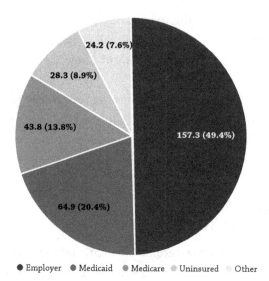

● Employer ● Medicaid ● Medicare ● Uninsured ○ Other

FIGURE 10.2   Breakdown of U.S. healthcare population by insurance providers.

Note: Total population 318.5 billion; total annual U.S. healthcare expenditure $3.3 trillion.
Source: Kaiser Family Foundation 2019.

Relative Value Unit and Resource-Based Relative Value Scale

By 1992, Medicare had implemented a **relative value unit** (**RVU**) system for physician practice expense payments. The RVU was designed to recognize differences in physician work, practice expense, and malpractice expense. RVUs are published annually as part of the **resource-based relative value scale** (**RBRVS**) developed by CMS. This system was designed to account for the resources consumed during the delivery of a physician's services. The **American Medical Association** (**AMA**) has a committee that provides recommendations of relative value to CMS; these recommendations are updated annually. This process is aimed at providing stability to the payment system while allowing physicians an opportunity to continuously improve it.

The Total RVU

There are multiple levels to determining the total RVU. The work component of an RVU will vary depending on physician skill and training and the intensity or productivity of the work. The practice expense addresses overhead costs. The third component incorporates malpractice expense factors and is generally weighted much less than the other two factors. The RVU can be modified for unique circumstances; this would be done on a case-by-case basis. The RVU is also adjusted for geographic location. A **geographic adjustment factor** (**GAF**) is applied to the RVU using a published **geographic practice cost index** (**GPCI**) that is listed for all locations (Cohen 2014). Today, the RVU also accounts for differences in facilities, and the formula for calculating the physician expense part of the RVU is treated differently depending on whether the service is performed in a hospital or in skilled nursing facility or another setting. Once the total RVU is calculated, it is multiplied by a conversion factor that Medicare publishes. The current value for one RVU is about $40.

Before you try to fully understand the complexities of the Kyphon case, let's look at a simpler scenario. Imagine that you are visiting a friend in New York City and dining at a fine steak house when you inadvertently cut your hand with your steak knife. Your hand is bleeding a lot, and you need stitches urgently. Luckily, your restaurant is right next to a big hospital, and you race over to the emergency room for care. The doctor who sutures your hand is a skilled plastic surgeon. Ironically, you recall that last year you had a similar situation when you were away on a fishing trip and cut your hand with your pocket knife. At that time, though, you went into the small nearby town, finding the local doctor in her office, where she sutured the wound.

When you receive your insurance statement for the steak knife incident, you compare it with the one you received last year. You see that even though you had very similar treatments in both cases, the cost for treatment in the New York hospital was much higher than that during your fishing trip!

This is because the RVUs for the New York episode are higher. The plastic surgeon had special skills and training. Perhaps if your wound was near a nerve or if you were worried about a scar, that surgeon would have more skills or training to address those concerns. The GAF and GPCI would be higher to reflect the higher cost of living in New York compared with a small town. The emergency room facility would warrant a higher rate factor than a doctor's office setting would—and this makes sense. It costs more to staff and equip an emergency room—it has to be ready for any scenario—than it does to staff and equip a doctor's office. By the way, please be more careful when handling knives in the future!

Important Design Knowledge

Make sure that your device has a code. It is important to note that a code must exist in order for an insurer to provide payment. If a claim is prepared without a code, payers will not reimburse, no matter how beneficial the service may have been. This becomes a very significant strategic consideration for a medical device designer. Often a new technology or device will include a new or alternate methodology for treatment. If this new method cannot be associated with an existing code, there will be little or no chance for payment until a code is created. It usually takes a couple of years—and maybe longer—to establish a new code.

In this period, the product and associated procedure would not generate income and would not attract the support of most investors or would have difficulty finding corporate champions. Despite the potential merit of the idea, if it is unable to generate revenue in a timely manner, it may not be commercially viable.

Summary

The Kyphon case demonstrates the complexity of the U.S. reimbursement system. In this case, the payment system was abused because patients were admitted to a hospital when they could have been treated as outpatients. But there are many cases where a similar treatment may appropriately be administered in either setting depending on other factors in the case. The story of Kyphon illustrates how payment methods are considered in the marketing of a device. A medical device designer must fully understand the device reimbursement scenarios so that an accurate business case can be established. More details about reimbursement codes and mechanisms are provided in Chapter 11.

Study Questions

1. How many fractures occur as a result of osteoporosis annually? What is the estimated annual cost to the U.S. healthcare system?
2. What is the difference between inpatient and outpatient status?
3. Explain how a kyphotic condition arises from vertebral fractures. Draw pictures if necessary.
4. What is the difference between a vertebroplasty and a kyphoplasty procedure?
5. What led to the adoption of employer-provided health insurance in the United States?
6. How are all insured people affected by Medicare's reimbursement methods?
7. In the early 1970s, there was concern about what? What did that concern lead Congress to do?
8. Who created RVUs and why? What are they created to do?

Thought Questions

1. Look up other spinal treatments for vertebral compression fractures. What do you think leads to kyphoplasty being more popular than many of those options?
2. What is a GAF? What could account for this difference, and do you think that it is appropriate/fair?
3. Why did Medtronic feel comfortable acquiring a company it knew was facing legal trouble? What do you think this strategy means for medical device designers and their patients?
4. Research the committee that makes suggestions to the AMA on relative values: the AMA Specialty Society RVS Update Committee (RUC). What qualifications must you have to be on the committee? What power and what limitations does the committee have?

References

Carroll, A. E. 2017. "The Real Reason the U.S. Has Employer-Sponsored Health Insurance." *New York Times*, September 5.

Centers for Medicare and Medicade Services (CMS). 2015. "Milestones 1937–2015." Baltimore, MD. www.cms.gov/About-CMS/Agency-Information/History/Downloads/Medicare-and-Medicaid-Milestones-1937-2015.pdf.

Cohen, F. 2014. "The Basics of Making RVUs Work for Your Medical Practice." *Physicians Practice*, July 1 www.physicianspractice.com/view/basics-making-rvus-work-your-medical-practice.

Halwani, T., and M. Takrouri. 2006. "Medical Laws and Ethics of Babylon as Read in Hammurabi's Code (History)." *Internet Journal of Law, Healthcare and Ethics* 4(2):1–8.

Jha, R. 2013. "Current Status of Percutaneous Vertebroplasty and Percutaneous Kyphoplasty: A Review." *Medical Science Monitor* 19:826–836. https://doi.org/10.12659/MSM.889479.

Kaiser Family Foundation. 2019. "Health Insurance Coverage of the Total Population." San Francisco. www.kff.org/other/state-indicator/total-population.

McCall, T., C. Cole, and A. Dailey. 2008. "Vertebroplasty and Kyphoplasty: A Comparative Review of Efficacy and Adverse Events." *Current Reviews in Musculoskeletal Medicine* 1(1): 17–23. https://doi.org/10.1007/s12178-007-9013-0.

Riggs, B. L., and L. J. Melton. 1995. "The Worldwide Problem of Osteoporosis: Insights Afforded by Epidemiology." *Bone* 17(5 Suppl. 1):S505–S511. https://doi.org/10.1016/8756-3282(95)00258-4.

Sherman, B. D., M. Chairman, R. Young, et al. 2008. "Medtronic to Buy Kyphon to Expand Spine Business." Reuters, July 27, https://www.nytimes.com/2017/09/05/upshot/the-real-reason-the-us-has-employer-sponsored-health-insurance.html.

Williams Walsh, M. 2008. "Medtronic Settles a Civil Lawsuit on Allegations of Medicare Fraud." *New York Times*, May 23, pp. 6–7.

CHAPTER 11

Navigating Codes for Reimbursement

Learning Objectives

Gain exposure to common reimbursement terminology.
Recognize where and how reimbursement codes are used.
Appreciate why cost/pricing strategies for medical devices are linked to reimbursement codes.

New Terms

Current Procedural Terminology (CPT)
International Classification of Diseases (ICD)
Diagnosis-Related Groups (DRGs)
Major Diagnostic Categories (MDCs)
Healthcare Common Procedure Coding System (HCPCS)
durable medical equipment, prosthetics, orthotics, and supplies (DMEPOS)

Current Procedural Terminology Codes

One of the most familiar and perhaps the most prevalent coding systems is the **Current Procedural Terminology (CPT)** system. These codes are maintained, revised, and updated by the American Medical Association (AMA). The CPT Editorial Panel is comprised of 17 members, including physicians who are nominated by the national medical specialty societies, physician representatives from the Centers for Medicare and Medicaid Services (CMS), major private payers, and the American Hospital Association. An external party such as a medical device designer may solicit a change, update, or new code by providing written application and justification to this panel. Adoption, however, is not imminent; the change request may be postponed or rejected. Ultimately, the AMA representatives assign or decline codes and determine the relative value of medical and surgical procedures and diagnostic services by virtue of their recommendations.

International Classification of Diseases Codes

While CPT codes provide specific procedural information, **International Classification of Diseases (ICD)** codes provide information about diagnoses. Often both ICD and CPT codes are used to describe the medical process for the patient, and both are required by payers. The ICD coding system does include some procedure codes, but these are for inpatients only and would be used to pay the hospital for inpatient care. As of October 2015, American healthcare systems have undergone a significant transition from the ninth (ICD-9) to the tenth edition (ICD-10) of the ICD; this transition was aimed at expanding the capacity and specificity of the coding system. Also, it updated the 40-year-old ICD-9 codes to align with current practice. For example, concepts such as the Glasgow coma scale and blood alcohol were not defined in ICD-9 (Centers for Disease Control and Prevention [CDC] 2015). This was a long, progressive change, and you may still encounter many references to ICD-9.

In June 2018, the World Health Organization (WHO), which developed the ICD system, introduced a preview of ICD-11. This version is designed to simplify the coding structure. In addition, it will be completely electronic for integration with electronic medical records and data sources. ICD-11 is expected to go into effect in 2022.

Diagnosis-Related Groups

In 1983, Medicare implemented another tactic to manage rising healthcare costs: the **Diagnosis-Related Groups (DRGs)**. Up to that time, payment was made based on the individual service or product provided to the patient. Length of patient stay and other resource usage varied widely between hospitals, often without clear justification for the variations. The purpose of the DRGs is to relate a hospital's case mix to its resource demands and associated costs.

The DRGs are a patient classification scheme. It aligns patients who consume similar hospital resources into classes. When you go to a large hospital, you will often find patients physically aligned in a similar manner. For example, you might find all cardiac patients on one floor or all eye patients in a certain ward. Similarly, the DRG system has classified patients into about 25 **Major Diagnostic Categories** (**MDCs**) that generally have similar resource intensity usage. These MDCs are organized by organ system (Table 11.1). This

TABLE 11.1   Major Diagnostic Categories in the DRG System

MDC	Description
1	Diseases and Disorders of the Nervous System
2	Diseases and Disorders of the Eye
3	Diseases and Disorders of the Ear, Nose, Mouth and Throat
4	Diseases and Disorders of the Respiratory System
5	Diseases and Disorders of the Circulatory System
6	Diseases and Disorders of the Digestive System
7	Diseases and Disorders of the Hepatobiliary System and Pancreas
8	Diseases and Disorders of the Musculoskeletal System and Connective Tissue
9	Diseases and Disorders of the Skin, Subcutaneous Tissue and Breast
10	Endocrine, Nutritional and Metabolic Diseases and Disorders
11	Diseases and Disorders of the Kidney and Urinary Tract
12	Diseases and Disorders of the Male Reproductive System
13	Diseases and Disorders of the Female Reproductive System
14	Pregnancy, Childbirth and the Puerperium
15	Newborns and Other Neonates with Conditions Originating in the Perinatal Period
16	Diseases and Disorders of the Blood and Blood Forming Organs and Immunological Disorders
17	Myeloproliferative Diseases and Disorders, and Poorly Differentiated Neoplasm
18	Infectious and Parasitic Diseases (Systemic or Unspecified Sites)
19	Mental Diseases and Disorders
20	Alcohol/Drug Use and Alcohol/Drug Induced Organic Mental Disorders
21	Injuries, Poisonings and Toxic Effects of Drugs
22	Burns
23	Factors Influencing Health Status and Other Contacts with Health Services
24	Multiple Significant Trauma
25	Human Immunodeficiency Virus Infections

is because even if patient conditions are different, the patients will use many of the same resources, and the intensity of use of those resources in the overall patient mix is expected to be similar (Centers for Medicare and Medicaid Services [CMS] 2016).

Important Design Knowledge

Be sure you understand the price that a hospital can charge for your device. If you are designing a device or technology for hospital use, you will surely encounter the concept of DRGs. What's ultimately important for you to consider is what price can the hospital charge for your device? Keep in mind that the DRG system "bundles" many products and services. Be certain that there is room within the DRG to pay for your device or procedure. Also, some technologies might fit into more than one MDC. In this case, it is important to understand the value proposition in each of these categories because value may vary from one classification to another.

Example of How DRGs Work

Suppose that two of your neighbors are admitted to your nearby hospital for treatment of the flu. The hospital kept old Mr. Curmudgeon for eight days but released young Mrs. Sweetwater after 24 hours. It doesn't matter how long or intensive the treatment actually is. Medicare has set standard fees for a DRG based on the cost expectations for a typical case of the flu. Because they had the same diagnosis, both Mr. Curmudgeon and Mrs. Sweetwater will be billed the same using the same DRG.

In actuality, because Mrs. Sweetwater recovered quickly, your hospital probably didn't spend as much as was allocated in the DRG, so it may have made a little extra money. In contrast, the hospital probably lost some money because it had to keep Mr. Curmudgeon for longer than normal. You should make sure that you get your flu shot, by the way.

Healthcare Common Procedure Coding System and Durable Medical Equipment, Prosthetics, Orthotics, and Supplies

If your device or equipment is intended for use as medical supplies, durable medical goods, or services from someone other than a physician and it is not covered by a CPT code, the associated billing code may come from level II of the **Healthcare Common Procedure Coding System (HCPCS)**, often pronounced "hick picks." This includes **durable medical equipment, prosthetics, orthotics, and supplies (DMEPOS)** when used outside a physician's office. You may also encounter the HCPCS coding system if your technology involves outpatient hospital care, drugs, infusion additives, ancillary surgical supplies, and ambulance services.

In 2006, CMS enacted safeguards to ensure that suppliers of DMEPOS meet certain quality standards. Suppliers must be accredited in order to enroll for or maintain Medicare billing privileges. The accreditation requirement applies to suppliers of durable medical equipment, medical supplies, home dialysis supplies and equipment, therapeutic shoes, parenteral/enteral nutrition systems and products, transfusion medicine and prosthetic devices, and prosthetics and orthotics (Medicare Learning Network 2015). If the device or service that you are designing falls into any of these categories, the market will be constrained by the suitability to accredited suppliers, and its marketing will be governed by those accredited suppliers.

When I was a kid, a little boy in my class had a "wooden leg." He had been hit by a car, and the prosthesis used to replace his lost lower limb was literally made of wood. It was fairly simple in design, and I remember the distinctive way that he skipped when he walked or ran, which became more pronounced as he grew. In all the time that I knew him, I don't recall him ever having the prosthetic adjusted or replaced. Thankfully, we are living in a period where prosthetics and orthotics are rapidly becoming more sophisticated. Technologies, materials, and engineering designs have evolved to meet socially prominent needs for disabled athletes and veterans and to support accessibility initiatives. Orthotics and prosthetics (O&Ps) are no longer as simple as durable medical equipment. These types of devices involve a much greater investment than they did when the DMEPOS category was established. Look for the O&P profession to ask to separate orthotics and prosthetics reimbursement structure from durable medical equipment in the near future.

Important Design Knowledge

Consider how payment influences your design. Let's return now to the Kyphon case that was presented in Chapter 10. Consider how the coding systems are different depending on the setting in which the procedure is performed. The payer of a patient whose kyphoplasty procedure was performed as an outpatient is billed using the resource-based relative value scale (RBRVS) methods discussed earlier. However, if that same patient were admitted for an overnight stay, the payer would be billed using DRGs. Because the kyphoplasty procedure is relatively simple compared with most back surgeries, the demand for resources likely would be quite a bit less than other surgeries in its DRG. But because the Kyphon procedure was categorized in a "high-resource-demanding DRG," the billing for use of equipment that falls into that DRG could be high. Ultimately, this would be a profitable proposition for a hospital; it could collect payment that was aimed at reimbursing it for resources needed to support a far more complex spinal surgery case.

Summary

While the courts declared that the Kyphon case represented a fraudulent abuse of the system, not all decisions regarding where and how to treat are as clear-cut. As long as there are financial incentives to drive a decision, there will be pressure to do so. Conversely, what is very important to appreciate as a device developer is that if the financial incentive is not in place, a good technology may never materialize commercially. Because payments are tied to this reimbursement structure, you have to consider these payment mechanisms for equipment, procedures, and services related to your device when evaluating the cost, price, and marketability of it.

Study Questions

1. What is the CPT coding system? Who has a say in how the codes change?
2. What is the ICD? What makes it different from the CPT? Do you need both for billing? Why?
3. What is the DRG system? Describe how DRG codes differ from CPT codes.
4. How does the text example of Mr. Curmudgeon and Mrs. Sweetwater relate to the Kyphon story?

Thought Questions

1. Revisit Table 11.1. Which of the categories do you think has the highest resource demand and why? If you had a product that fit into two different DRG's, to which one would you assign your medical device?
2. What justification could a medical professional make for keeping someone overnight for a kyphoplasty procedure? How might that have contributed to the Kyphon case?
3. Why might modern O&P professionals want to separate from the DMEPOS category?
4. Financial incentive is a massive driver in device design. Where else (other devices, other fields) have you seen financial incentives drive innovative and rapid product development? Where have you seen this system lead to neglected causes?

References

Centers for Disease Control and Prevention (CDC). 2015. "International Classification of Diseases, (ICD-10-CM/PCS) Transition: Background." Washington, DC, October 1, pp. 9–10.

Centers for Medicare and Medicaid Services (CMS). 2016. "Design and Development of the Diagnosis Related Groups (DRG)." Baltimore, MD, pp. 1–14.

Medicare Learning Network. 2015. "DMEPOS Accreditation."

CHAPTER 12

Device-Associated Infections

Learning Objectives

Learn about the healthcare-associated infections (HAIs).
Appreciate the costs and risks associated with HAIs.
Grasp why infection control must be considered in the design of a medical device.
Learn about infection control methods that are used in medical device design.
Appreciate the challenges that may develop when designing a device with characteristics aimed at infection prevention.

New Terms

total knee arthroplasty (TKA)
nosocomial infection
healthcare-associated infection (HAI) or hospital-acquired infection (HAI)
surgical site infection (SSI)
indication
revision
biofilm
indwelling
urinary tract infections (UTIs)
broad spectrum
infective endocarditis (IE)
prosthetic valve endocarditis (PVE)
paravalvular regurgitation (PVR)
unintended consequence

Nosocomial Infection (aka HAI)

A patient is admitted to a hospital for a knee replacement, otherwise termed a **total knee arthroplasty** (**TKA**). However, as the patient is recovering from surgery in the hospital, it is noticed that he has acquired an infection around the implant at the site of the surgery. This is known as a **nosocomial infection**: an infection acquired in a healthcare setting.

A nosocomial infection can also be called a **healthcare-associated infection** (**HAI**) or a **hospital-acquired infection** (**HAI**). This is an infection contracted in a healthcare-providing location such as a hospital (Stubblefield 2017). People who get nosocomial infections usually spend 2.5 times longer in the hospital and can be placed at much higher risk for life-threatening scenarios. (Inweregbu, Dave, and Pittard 2005). By some estimations, half of all HAIs are related to or affected by medical devices, and for this reason, HAIs are worth reviewing (Lillis 2015).

A **surgical site infection** (**SSI**), such as the one acquired by our knee arthroplasty patient, is a major type of HAI. In addition to SSIs, other major types of HAIs include central line–associated bloodstream infection (CLABSI); catheter-associated urinary tract infection (CAUTI); methicillin-resistant *Staphylococcus aureus* bacteremia (MRSA bacteremia); *Clostridium difficile* infection (CDI), often referred to as "C-diff"; and ventilator-associated pneumonia (VAP) (Centers for Disease Control and Prevention [CDC] 2017).

HAIs are significant and costly because their prevalence is associated with a high degree of morbidity and mortality. The CDC reports that on any given day, about 1 in 25 hospital patients has at least one HAI. It is estimated that around 800,000 people contract HAIs annually in the United States and that nearly 10 percent of those cases (or about 75,000 people) result in death (CDC 2017). Costs can exceed $90,000 per infection when an HAI involves a prosthetic joint implant or an antimicrobial-resistant organism, as in the case of SSIs (Berriós-Torres et al. 2017a). Despite recent significant reductions in the incidence of HAIs, the cost burden remains staggering. The overall direct medical costs of HAIs to U.S. hospitals are estimated to be in the range of $30 billion to $45 billion a year (Scott 2009).

Preventing HAIs from Medical Devices

Hand washing is the single most important measure to prevent nosocomial infections (Inweregbu, Dave, and Pittard 2005). In fact, if best practices for infection prevention and control such as hand washing and covering sneezes were applied consistently and effectively at all U.S. hospitals, they could reduce the risk of infection by as much as 70 percent (Umscheid et al. 2011).

The market for medical devices such as prosthetic joints and heart values is increasing as the world's population ages. It is expected that by 2030, there will be 3.8 million prosthetic joint arthroplasty procedures per year, and with this increase, the number of hip and knee joint infections is projected to increase to 221,500 cases per year in the United States. This has an estimated cost of more than $1.62 billion annually (Berriós-Torres et al. 2017).

Infection can impede tissue healing around implants, preventing the tissue surrounding the device from integrating with the device as needed. Such a complication can lead to loosening of the joint prosthesis or device, causing it to not work properly. Infections such as these can create the need to revisit or even remove an implant, which is costly and dangerous.

Actually, infection is the most common **indication** for revision in TKA and the third most common indication in total hip arthroplasty (Berriós-Torres et al. 2017). This **revision** of the implant, as well as the treatment of the infection, and the overhead cost of a patient being in the hospital longer are what contributes to this high cost.

Biofilms and Implant Failure

Preventing infection is paramount to the efficacy of a medical device. One of the most significant factors in the development of a medical device–associated infection is the formation of a bacterial **biofilm** that adheres to device surfaces. Such infections can be life threatening, and frequently, antimicrobial therapy fails unless the implanted device is removed. Therefore, the prevention of a biofilm is crucial.

A biofilm is a colony of microbial cells enclosed in a matrix of mostly polysaccharide material that adheres to a surface and is typically quite slippery. An analogy is a rock that has been underwater in a stream for a long period of time. If you try to step on this rock to cross the stream, you would find it to be very slimy. Not all biofilms are bad. Beneficial biofilms do exist in our bodies—for example, a biofilm in our intestines prevents potentially harmful bacteria from entering, attaching to, and invading our tissue.

Biofilms may form on a wide variety of surfaces, including living tissues and **indwelling** medical devices (Donlan 2002). A biofilm colony can promote increased resistance to biocides, antibiotics, and the immune system of the host (Lejeune 2010). A biofilm can be 1,500 times more resistant to an antibiotic than the same bacteria in free-floating form (Gibbins and Warner 2005).

Because of the highly resistant nature of biofilms, the effectiveness of oral antibiotics or other antimicrobial strategies can be severely compromised. In such cases, the implant may need to be removed and the infection site cleaned, possibly with another surgical procedure. The infection would then need to fully resolve before the patient could be considered for a new implant in a **revision** surgery. The prior infection also creates a higher level of risk for that revision surgery.

Sometimes a bacterial colony can adversely affect the function of an implant even before a clinical infection is evident. Some common examples include the failure and required removal of mammary implants, biliary stents, and dental implants. Similarly, a biofilm can cause vascular occlusion, as in the case of a central venous catheter (Darouiche 2001).

Device-Related Infections and Urinary Tract Infections

Indwelling medical devices are highly susceptible to the formation of a biofilm partly because of their negative biocompatibility with the host and often positive biocompatibility with microbial cells that adhere well to the device surface. As a result, many HAIs are considered to be device related. For example, it is estimated that 80 percent of **urinary tract infections (UTIs)** are related to urinary catheters (CDC 2017).

Urinary catheterization is a common procedure for many patients who are hospitalized. Between 10 and 30 percent of patients who undergo short-term catheterization and more than 90 percent who are catheterized longer term will develop bacteria in their urine. While this will not always lead to an infection, it predisposes a patient to infection risk because bacteria may enter into the bladder during insertion or removal of the catheter or even during manipulation of the catheter or of the drainage system around it. Urethral catheters can promote bacteria colonization by causing mucosal irritation on insertion, manipulation, and removal. Also, even though they are indwelling, these catheters provide a surface for bacterial adhesion (Brusch 2017).

Catheter-associated UTIs (CAUTIs) account for more than 30 percent of HAIs and are the most common type of HAI. The average cost of a CAUTI is approximately $750, which is relatively low compared with SSIs (described earlier). However, there are approximately 450,000 CAUTI events in the United States each year, with 13,000 of these (nearly 3 percent) resulting in death annually. The overall cost burden is still quite large because of the high frequency and volume of annual cases. If you do the math here, you can see that more than $340 million spent on healthcare in the United States each year is attributable to CAUTIs (CDC 2017).

Surface Treatment to Prevent Infection

So far we have discussed how biofilm formation on the surfaces of indwelling devices can lead to infection. You might imagine that a good strategy would be to somehow treat the surface or otherwise adapt the device materials to prevent biofilm formation or to kill microbes as they land on the surface. Indeed, this is what medical device manufactures have strived to do in recent years.

The Healthcare Infection Control Practices Advisory Committee (HICPAC), a federal advisory committee (CDC 2020), has prepared guidelines for the prevention of CAUTIs, some of them materials based. These guidelines include considering the use of antimicrobial/ antiseptic-impregnated catheters, although the HICPAC identifies the need for further research to understand their efficacy.

HICPAC also suggests the use of hydrophilic catheters. A hydrophilic catheter is one that is surface treated with a coating that becomes slippery, or lubricious, when wet. HICPAC also suggests that an alternate material, silicone, might be preferable to other catheter materials to reduce the risk of encrustation due to its inherent slipperiness when wet. The main benefits of a lubricious surface are twofold. First, the surface of the device will be less irritating to the surrounding tissue, thus reducing inflammation. Second, the slippery surface reduces the opportunity for microbial cells to adhere to the surface, making it harder for them to colonize.

Challenges for Antimicrobial Surface Strategies

Another option for handling infections is antibiotics. The dosage and duration of antibiotic treatment must be carefully considered because if the treatment is not sufficient, the bacteria have time and opportunity to develop resistance. This not only renders the antibiotic useless in treating that infection but also builds a "new and improved" bacterial strain that is even more resistant than the original. Antimicrobial resistance can be deadly and makes the infection more difficult to beat with the drugs that are currently available. Overuse of an antibiotic can also lead to antimicrobial resistance, as well as cause serious, or even deadly, side effects.

Because of the number of different bacteria that can cause harm in the human body, any medical device would need the antimicrobial agent impregnating its surface to be **broad spectrum**—affecting multiple strains of bacteria. Furthermore, it is important to note that the characteristics of pathogens mutate when they change from their free-floating state to their biofilm form, and for this reason, too, a broad-spectrum agent is needed.

An antimicrobial agent for a device surface ideally should be hostile to the colonization of biofilms and be capable of broad-spectrum resistance. In the best of cases, these properties would not be diluted or washed away by blood and bodily fluids when the device is implanted. Similarly, the agent should adhere well enough that it does not wear off by abrasion on insertion or while indwelling. Ultimately, the agent must be in a form that does not interfere with the efficacy of the device but that does allow for effective antimicrobial activity.

Silver as an Antimicrobial Agent

Silver is one of the longest known antimicrobial agents. In the right form, it has been proven that many bacteria and other microbes are not resistant to it. This is a result of the dual nature by which it attacks, affecting both the membranes of microbes and their ability to reproduce. Meanwhile, it has a low affinity for human tissue, making it relatively safe and noninflammatory. Silver has been increasingly looked upon as a favored substance for surface modification of medical devices. Additional considerations for the preference of silver over other materials are that silver is compatible with most materials used in making medical devices and provides for the option of being applied either to the surface or compounded directly into the device material (Gibbins and Warner 2005).

Silver and Urinary Catheters

Given the high incidence of CAUTIs and the promising properties of silver as an anti-infective agent, it would make sense to consider the use of silver on urinary catheters. In fact, two large medical device companies, Bard and Tyco (now Covidien), successfully introduced silver-coated urinary catheters near the turn of the millennium. Since then, several studies have been performed to evaluate the effectiveness of these catheters. Although the quality of many of these studies is weak, there is some indication that for short-term catheterizations, silver-coated urinary catheters may be beneficial in the prevention of UTIs (Ford, Hughes, and Phillips 2016).

Infective Endocarditis and Prosthetic Valve Endocarditis

Another type of device that could benefit from an effective surface treatment strategy is the prosthetic heart valve. **Infective endocarditis (IE)** is an infection of the heart. **Prosthetic valve endocarditis (PVE)** accounts for about one-fifth of all cases of IE. Between 1 and 4 percent of patients who receive a heart valve replacement will suffer from PVE. While the incidence is low, it is a very serious condition and results in death about 20 percent of the time. In many cases, the implant must be removed and replaced in a revision surgery. In fact, approximately half the patients who contract PVE undergo surgical treatment. Imagine how much risk and expense are involved in removing and replacing a heart valve.

Silzone by St. Jude Medical

Recognizing the significance of IE and PVE, one of the largest makers of prosthetic heart valves, St. Jude Medical (SJM), introduced a prosthesis in 1997 that included silver. In an attempt to reduce the incidence of PVE, the company added a silver-coating technology on the sewing cuff of its conventional mechanical heart valves (Figure 12.1). This silver product was called Silzone, which the company describes as a thin application of elemental silver. The idea was to inhibit colonies from developing at the site where the device and sutures connect with the host tissue (Baghai et al. 2007).

FIGURE 12.1   The St. Jude "Silzone" Artificial Heart Valve with a Silzone-coated white sewing ring or cuff that was sewed directly to the patient.

The Artificial Valve Endocarditis Reduction Trial and Paravalvular Regurgitation

To study the efficacy of its Silzone coating, St. Jude Medical established a clinical trial, known as the Artificial Valve Endocarditis Reduction Trial (AVERT). It was designed to evaluate how effective the silver-coated sewing cuff of the company's mechanical heart valve was in reducing infection following valve replacement surgery. It was planned as the largest, most rigorous prospective clinical trial ever in the prosthetic heart valve industry and was intended to follow 4,400 patients.

By early January 2000, data from approximately 800 patients enrolled in AVERT had been reviewed, and it was observed that an unacceptable level of explants had been required in the Silzone arm of AVERT because of paravalvular leakage. Paravalvular leakage, also termed **paravalvular regurgitation** (**PVR**), occurs when blood leaks at one or more points around the outside of the implanted valve, between the sewing cuff of the valve and the heart tissue. Major paravalvular leaks were observed in 18 (4.4 percent) of 403 patients receiving Silzone-coated valves and in 4 (1.0 percent) of 404 patients receiving valves without the Silzone coating (Baghai et al. 2007). Eight patients enrolled in the Silzone arm of the study had PVR severe enough that they required explants compared to only one patient in the control arm, where valves without the Silzone coating were used.

Later that same month, an independent data and safety monitoring board advised St. Jude Medical to suspend enrollment in the AVERT study, which the company heeded. Furthermore, as a precaution, St. Jude recalled and ceased distribution of tissue valve and repair products with the Silzone coating. Still, by the time St. Jude ceased distribution of the Silzone heart prosthesis, an estimate 36,000 patients had received the valve. Since the AVERT study, other studies have been performed, but none have been able to deliver definitive findings (Baghai et al. 2007).

As a device designer, you might wonder if the Silzone coating that was impregnated in the fabric of the valve's sewing cuff affected the design, manufacture, or function of the valve mechanism. St. Jude Medical determined that this was not the case. The company indicated that PVR could occur with or without an infection. On inspection of an uninfected case, however, it was observed that the sutures of the sewing ring failed to remain firmly anchored to the tissues of the heart. Thus, whether an infection occurred or not, it appears that the Silzone sewing ring interfered with the security of the sutures, leading to PVR. The nature of the mechanism that caused this complication remains in question. One prominent theory suggests that the Silzone coating inhibits normal fibroblast growth into the prosthetic sewing cuff, impairing tissue ingrowth and leading to loosening of sutures (Baghai et al. 2007). Loose sutures allow for leakage—PVR.

Whereas Silzone treatment of the mechanical heart valve was intended to decrease the incidence of infection, there appears to have been an **unintended consequence**. In addition to preventing microbe development, Silzone appears to have also inhibited endothelial development on the sewing ring of the valve, thus preventing integration of the host tissue with the implant. In other words, the silver coating seems to have been indiscriminate, and although it may have prevented bad microbes from colonizing, it also prevented good cells from attaching to where they were needed for the implant to remain securely seated. Unintended consequences are not uncommon in medicine. There are many factors that contribute to the success or failure of a treatment, and those factors span across the disciplines of biology, engineering, clinical practice, and manufacturing.

Even with a well-thought-out design, such as the Silzone-coated valve, outcomes might occasionally be surprising or detrimental. This is why it is so important to consider the historical perspective of similar approaches and to follow prescribed medical design practices. These practices, while always a work in progress, have been developed based on past experiences and established to account for multifactorial influences. St. Jude Medical did all the right things when it introduced Silzone, and the company could not have anticipated the high rate of PVRs. But we can learn from the company's experience when considering the use of silver in medical devices and in general when considering antimicrobial strategies.

Important Design Knowledge

In addition to surface treatments, there are many other device-related factors that may favor bacterial adherence, such as device materials, surface textures, and shapes. For example, bacteria are more likely to adhere to stainless steel than to titanium, and in general, a natural biomaterial is less likely to encourage bacterial adherence than a synthetic material (Darouiche 2001).

When designing an infection-prevention strategy for an implant, it is very important to consider any related experiences from previous device failures. It is also critical to recognize that the application of an infection-prevention strategy is very specific to that particular usage, and you must carefully identify and evaluate the differences in addition to the similarities. As shown in the preceding examples, silver has shown some promise for infection prevention as a surface treatment on urinary catheters, but it was catastrophic on a heart valve prosthesis. Most important, you must recognize that microbes that cause infections are living organisms that can change their properties; they can strengthen when colonized, and given the opportunity, they can adapt to grow more resistant to known strategies.

Summary

Hospital acquired infections are unfortunately all too common. While there have been significant reductions in these infections, the health and cost implications for patients and hospitals are significant. Infections can be introduced by a variety of ways but over half of all HAIs are related to medical devices. Urinary catherization is a common procedure for many patients who are hospitalized and it is estimated that 80 percent of UTI's are related to urinary catheters.

Take my friend, Steve. While healthy and active, he needed a minor inpatient procedure. Shortly after he got home, he developed a rash. He went back to the doctor to have it checked. He had contracted a MRSA infection and was re-admitted to the hospital for treatment. He woke up two months later, lucky to be one of the survivors. While he was unconscious for all that time, he had a urinary catheter. Despite a successfully waning MRSA infection, he acquired a UTI. He is home now but it will be several months before he can work or be active again. These infections caused substantial implications to both him and the hospital.

Study Questions

1. Define *nosocomial*. What is the etymology of this word?
2. How does nosocomial infection affect a person's length of hospital stay and level of risk?
3. What is the average cost per case of a prosthetic joint infection?
4. How much money could we save in the United States if hand washing and other best practices for preventing nosocomial infections are adhered to strictly?
5. Why would an oral antibiotic potentially be a good strategy for treating a UTI?
6. What type of device surface do you think favors bacterial adherence?
   - Irregular or regular?
   - Textured or smooth?
   - Hydrophobic or hydrophilic?

Thought Questions

1. Apply the results from the AVERT study to the general population. How many patients may have suffered from PVR unnecessarily because of the Silzone coating on the St. Jude Medical Artificial Heart Valve?
2. Research an example of a beneficial biofilm and a harmful biofilm not mentioned in this chapter. What engineering strategies have been used to either cultivate or curtail this growth?
3. Describe the unintended consequence from the use of silver on the Slizone-coated heart valve ring. Explain why you think this occurred.
4. Which would you be more likely to green light as a Food and Drug Administration regulator based on the suspected issues with the Silzone-coated heart valve, a silver-impregnanted vertebral bone graft or a silver-impregnated stent?
5. Read a news article from the last four months on modern antibiotic resistance. What do you predict will happen in the world of nosocomial infections in the next five years? How could silver play a role?

References

Baghai, M., U. P. Dandekar, M. Kalkat, and P. D. Ridley. 2007. "Silzone-Coated St. Jude Medical Valves: Six-Year Experience in 46 Patients." *Journal of Heart Valve Disease* 16:37–41. www.researchgate.net/publication/6491300.

Berriós-Torres, S. I., C. A. Umscheid, D. W. Bratzler, et al. 2017. "Centers for Disease Control and Prevention Guideline for the Prevention of Surgical Site Infection, 2017." *JAMA Surgery* 152(8):784–791. https://doi.org/10.1001/jamasurg.2017.0904.

Brusch, J. L. 2017. "Catheter-Related Urinary Tract Infection (UTI)." Medscape, September 8. https://emedicine.medscape.com/article/2040035-printemedicine.medscape.com.

Centers for Disease Control and Prevention (CDC). 2017. "Data Summary of HAIs in the US: Assessing Progress 2006–2016." HAI Data Report, Washington, DC, October 5. www.cdc.gov/hai/surveillance/data-reports/data-summary-assessing-progress.html.

———. 2020. "Healthcare Infection Control Practices Advisory Committee (HICPAC)." Washington, DC, September 25. www.cdc.gov/hicpac/index.html.

Darouiche, R. O. 2001. "Device-Associated Infections: A Macroproblem That Starts with Microadherence." *Clinical Infectious Diseases* 33(9):1567–1572. https://doi.org/10.1086/323130.

Donlan, R. M. 2002. "Biofilms: Microbial Life on Surfaces." *Emerging Infectious Diseases* 8(9):881–890.

Ford, J., G. Hughes, and P. Phillips. 2016. "Literature Review of Silver-Coated Urinary Catheters." Medidex: Surgical Materials Testing Laboratory, July, pp. 1–23. www.medidex.com/research/830-silver-coated-catheters-full-article.html.

Gibbins, B., and L. Warner. 2005. "The Role of Antimicrobial Silver Nanotechnology. *Medical Device and Diagnostic Industry Magazine*, August, pp. 1–6. www.acrymed.com/pdf/Antimicrobial Silver Prevents Biofilm Formation-MDDI 8-2005.pdf.

Inweregbu, K., J. Dave, and A. Pittard. 2005. "Nosocomial Infections." *Continuing Education in Anaesthesia, Critical Care & Pain* 5(1):14–17. https://doi.org/10.1093/bjaceaccp/mki006.

Lejeune, P. 2010. "Biofilm-Dependant Regulation of Gene Expression." In M. Wilson and D. Devine (eds.), *Medical Implications of Biofilms* (4th ed., vol. 1, pp. 3–17). Cambridge, UK: Cambridge University Press. https://doi.org/10.1016/B978-012095440-7/50028-7.

Lillis, K. 2015. "Device-Associated Infections: Evidence-Based Practice Remains the Best Way to Decrease HAIs." *Infection Control Today*, April 11. www.infectioncontroltoday.com/view/device-associated-infections-evidence-based-practice-remains-best-way-decrease-hais.

Scott, R. D., II. 2009. "The Direct Medical Costs of Healthcare-Associated Infections in U.S. Hospitals and the Benefits of Prevention." Division of Healthcare Quality Promotion, National Center for Preparedness, Detection, and Control of Infectious Diseases, Coordinating Center for Infectious Diseases, Bethesda, MD. www.cdc.gov/hai/pdfs/hai/scott_costpaper.pdf.

Stubblefield, H. 2017. "What Are Nosocomial Infections?" *Healthline*, June 6 www.healthline .com/health/hospital-acquired-nosocomial-infections.

Umscheid, C. A., M. D. Mitchell, J. A. Doshi, et al. 2011. "Estimating the Proportion of Healthcare-Associated Infections That Are Reasonably Preventable and the Related Mortality and Costs." *Infection Control and Hospital Epidemiology* 32(2):101–114. https:// doi.org/10.1086/657912.

CHAPTER 13

Designing for Postmarket Safety

Learning Objectives

Learn about medical device cases where there were postmarket safety problems: Essure, spinal cord stimulators, surgical and transvaginal mesh.

Understand how long-range data can change a device's risk profile (balance must be struck between timely access to market and the generation of data from clinical studies of sufficient duration).

Learn about challenges with imaging to identify the etiology of back pain.

Understand that when assessing the potential value and impact of a medical device, it is critical to fully evaluate all possible solutions, including strategies that exist beyond the form of the medical device.

Appreciate the magnitude and impact of the opioid crisis and recognize the need for alternative solutions to treat pain.

Recognize how substantial safety risk may be introduced into the medical device marketplace via the 510(k) process.

Learn about pelvic floor disorders.

Appreciate how the Food and Drug Administration is asserting more authority, including market restriction device reclassification, market restriction, and forced market withdrawal.

New Terms

tubal ligation	spinal cord stimulator (SCS)
adverse event	urogynecologic mesh
magnetic resonance imaging (MRI)	pelvic floor disorders
discogenic back pain	pelvic organ prolapse (POP)
transcutaneous electrical nerve stimulation (TENS)	pessary
epidural steroids	colpocleisis
opioid crisis	adjuvant

Long-Range Data Come Only After a Product Is on the Market

Essure

Recall the discussion about the Dalkon Shield in Chapter 5. It was an intrauterine device (IUD) that was introduced in 1966 and that the Food and Drug Administration (FDA) ordered to be removed from the U.S. market in 1974. The device was available globally outside the United States until 1980. It caused injury to hundreds of thousands of women and led to the Medical Device Acts of 1976 (see Chapter 3).

In the wake of the disastrous results, a need for safe and effective contraception alternatives remained. It wasn't long before medical device manufacturers tried, once again, to introduce a safe and effective birth control to U.S. markets. By 2002, Conceptus, Inc., introduced the Essure product.

Essure was intended to be a permanent sterilization method and an alternative to **tubal ligation**. Tubal ligation is a surgical procedure in which the fallopian tubes are removed or blocked so that eggs cannot be fertilized. Essure was seen to have advantages over the surgical procedure; it could be implanted in a doctor's office. Concepts for the materials and placement technology were borrowed from stent manufacturers. The minimally invasive technologies used in interventional cardiology seemed to translate well to the insertion techniques of the Essure device. Short-term studies indicated that Essure appeared safe and effective and that it could be a less expensive and less invasive alternative to surgery.

By 2013, the Essure device had been placed in 750,000 women, and in June of that year, Bayer acquired Conceptus. Meanwhile, **adverse event** reports about Essure were mounting within the FDA. Complaints included persistent pain, hemorrhaging, headaches, perforation of the uterus or fallopian tubes from device migration, abnormal bleeding, and allergy or hypersensitivity reactions. Almost 1,000 complaints had been received. Information describing the possibility of unsuccessful device placement began to emerge (Deardorff 2013).

FDA Monitoring and Guidance

The FDA had been closely monitoring Essure since it was approved in 2002. The agency stated that it had reviewed the medical literature, clinical trial information, postapproval study data, and medical device reports that it received. In February 2016, the FDA announced that it would issue guidance that aimed to increase patient and physician understanding of the potential risks associated with this type of device.

The actions in the guidance included a requirement for a mandatory label warning on the box of the product "explaining the adverse events that have been associated with these devices, including their insertion and/or removal procedures." It also proposed language for a "patient decision checklist" (U.S. Food and Drug Administration [FDA] 2016b). This was intended to facilitate the communication of risks and to ensure an informed decision-making process for patients. The FDA also ordered Bayer to conduct a new postmarket surveillance study aimed at providing device risk information in a real-world environment (FDA 2016b).

FDA Invokes Rare Restrictions

Sales of Essure dropped by 70 percent after these actions were ordered. Still, in spite of the efforts to ensure that women were educated about the risks associated with the device, the FDA became aware that some women were still not being adequately informed before having the device implanted. Thus, on April 9, 2018, the FDA invoked a restriction on the sale and distribution of Essure in the United States. It is rare that the FDA exercises such authority, and this action implies that it viewed that the product had more risk than originally identified in the U.S. market approval process. On December 31, 2018, Bayer stopped selling and distributing the Essure device in the United States, with a caveat that healthcare providers can implant Essure up to one year from the date the device was purchased (FDA 2020).

Important Design Knowledge

Relative to the case of the Dalkon Shield 50 years earlier, in the Essure case, the FDA had significantly more authority to monitor and control the sale and distribution of medical devices. Still, at least 750,000 women had received the device. Short-term studies did not provide enough accurate information about the risks. The FDA continues to learn from these lessons, and short-term data may be highly scrutinized or discounted as sufficient evidence in future device reviews. A reasonable balance must be struck between timely access to the market and the generation of data from clinical studies of sufficient duration.

Back Pain and Pain Management

If you have never experienced back pain, you are in a very fortunate minority. Approximately 80 percent of Americans will experience back pain in their adult lifetime. Back pain is so prevalent that more than 25 percent of people have experienced back pain in the past three months, and it is a leading cause of job-related disability and work days lost (National Institute of Neurological Disorders and Stroke 2018). Recall from Chapter 1 that back problems are the seventh most expensive condition treated in hospitals, accounting for more than $10 million annually in hospital costs alone.

Most people who experience back pain feel it in their lower back. Often the pain is acute and resolves with self-care. Usually, back pain is due to a mechanical issue: some disruption of the way the spine, muscles, or nerves fit and move within the back. Common risk factors include poor fitness, age (see Chapter 10), weight gain, and jobs that require heavy lifting. However, if pain persists longer than 12 weeks, it is considered chronic, and further treatment is required (National Institute of Neurological Disorders and Stroke 2018).

Variability in Image Interpretation

To diagnose back pain, images are usually taken. In the early phases of back pain, the first course of images is usually x-rays. X-rays imaging will reveal the bony structures and is a good tool for visualizing fractures or other misalignment of bony tissue such as the vertebral bodies. However, the soft tissues, such as the discs, muscles, and ligaments, are not visible in an x-ray. If a doctor feels that there is something in the medical history or evaluation that warrants further imaging, he or she will request **magnetic resonance imaging** (**MRI**). An MRI image will add a view of the soft tissues, including the vertebral discs.

Even with the additional information from MRI images, their interpretation may depend significantly on which radiologist reads the results. In a recent study, a 63-year-old woman went to 10 different MRI centers within a three-week period and received a lumbar spine MRI. While there was a total of 49 findings identified in all the reports, there was not one finding in common among all 10 centers! There was only one finding in common among 9 of the 10 centers, and 33 percent of the 49 findings appeared just once across all the reports (Herzog et al. 2017).

This poor agreement among reviews should concern anyone seeking treatment. Furthermore, report findings on an MRI image have an impact on insurance coverage and even work status, and the treatment selected as a result of the findings will affect clinical outcomes.

The fundamental pathophysiology and diagnostic criteria of back pain are still debated. This contributes to the variability among reported interpretations of findings. This debate often extends to a comparison of surgical versus nonsurgical treatments. Clearly, then, more information is needed to understand and treat back pain.

In a community study performed in 2013, patients who underwent surgical and nonsurgical methods of treatment for **discogenic back pain** were monitored (Mirza et al. 2013). At one-year follow-up, the success rate was 33 percent for surgery and 15 percent for nonsurgical treatment. While the surgical group showed greater improvement at one year compared with the nonsurgical group, it is more significant to note that the composite success rate for both treatment groups was only fair.

Important Design Knowledge

There is significant room for the improvement of clinical outcomes for back pain. While researchers aim to better understand the etiology of back pain and the clinical community strives for improved diagnosis and treatment, there are many opportunities to improve technology to support this quest. Engineers continue to improve the capabilities of existing diagnostic and treatment tools, including x-ray, MRI, ultrasound, computed tomographic (CT) scans, electromyelography, surgical hardware, and the like. Engineers who can collaborate with the research and clinical disciplines to help bridge the needs and understanding offer the best chances for improved outcomes.

Drugs versus Surgery versus Device versus Conservative Care

As seen in the case of back pain, there may be different types of strategies to achieve the same outcome. It is almost always most sensible to begin treatment with conservative care unless the delay of a more invasive treatment poses a risk of further injury or other complications. Beyond the reasonable progression of interventional severity, however, there are often alternate types of treatments available, and the decision regarding which strategy to choose may vary greatly depending on the patient and his or her circumstances.

In a 2013 study, of the 495 patients enrolled, only 86 had surgery, whereas 409 opted for a nonsurgical approach (Mirza et al. 2013). Nonsurgical conservative methods might include diet, exercise, physical therapy, or spinal manipulation. These methods might also include augmentation with medical devices such as **transcutaneous electrical nerve stimulation (TENS)** or other electrical stimulation. Patients may be treated with nerve blocks or **epidural steroids**. Drug therapies, both over the counter and prescription, also may be used. Practically speaking, these various types of therapies are often combined both among those who received surgical treatment and among those who did not.

If the clinical outcomes for back pain surgery were better, and if the risks and costs were lower, it is likely that much more of this study's population would have opted for surgical intervention as opposed to the only 20 percent who did.

Important Design Knowledge

It is quite probable that medical device engineers will help to improve clinical outcomes as they work to advance the state of diagnostic tools and surgical hardware. Note, however, that if you viewed solutions to back pain only through the lens of a medical device engineer, you would miss the alternative solutions available in the spaces of drugs and other therapies, which may be competitive or complementary. When assessing the potential value and impact of a medical device, it is critical to fully evaluate all possible solutions, including strategies that exist beyond the form of a medical device.

Opioids

Unfortunately, some patients continue to suffer severe, debilitating pain, even after being treated with the best arsenal of surgical and nonsurgical interventions. For those patients, doctors began prescribing opioids. Opioids were used primarily to treat cancer pain at first. By 1990, the prescription of opioids for the treatment of non-cancer-related pain had increased sharply. Pharmaceutical companies played a significant role in this trend, both by providing reassurances that the risk of addiction to prescription opioids is low and by promoting their use for non-cancer-related pain. By 1991, deaths involving opioids began to rise, and by 1999, 86 percent of opioids were being used by patients with non-cancer-related pain (Liu, Pei, and Soto 2018).

These days, an average of 130 Americans die from an opioid overdose every day. Over the past 20 years, opioids have accounted for more than 770,000 deaths. While many of these deaths were the direct result of overprescription of opioids, others were the result of heroin and fentanyl abuse. Former prescription opioid users turned to these for less expensive and easier to obtain pain relief products. The impact of this situation is so large that the **opioid crisis** is now of comparable magnitude to the flu pandemic of 1918 and the AIDS crisis of the latter twentieth century (Healton, Pack, and Galea 2019).

Nowadays, we know that anyone who takes a prescription opioid can become addicted to it. The risk of addiction may be as high as one in four for patients receiving long-term opioid therapy (Centers for Disease Control and Prevention [CDC] 2020). There is mounting evidence that some drug manufacturers promoted their opioids with misleading claims about the likelihood of addiction and safe dosages. Others allegedly failed to report excessive deliveries to individual distributors or fraudulent prescriptions written by those suspected of diverting the drugs for illegal use. In March 2019, in what is sure to be the first of many lawsuits, the state of Oklahoma settled a suit against Purdue Pharma for $270 million. In a separate but related case, an Oklahoma judge ordered Johnson & Johnson to pay $572 million in damages and specified how the funds should be spent as reparations to victims (Healton, Pack, and Galea 2019).

Many opponents to these rulings are reluctant to accept this agreement because the value of both the damages and the company's revenues significantly exceed the settlement price. Overall, the opioid industry's global annual revenues exceed $25 billion (Healton, Pack, and Galea 2019). Proponents of the settlement feel that this quick settlement will provide for an early and much needed use of the funds.

Spinal Cord Stimulators and Shortcomings of the 510(k) Process

"To reverse this [opioid] epidemic, we need to improve the way we treat pain. We must prevent abuse, addiction, and overdose before they start" (CDC 2020).

Here is a tremendous opportunity for a medical device. Eliminate the need for drugs! In the case of back pain, could a **spinal cord stimulator** (**SCS**) mitigate pain so that opioids are no longer needed? Well, yes and no. Up to 30 percent of SCS patients fail to reach long-term pain relief, even patients whose condition is considered to be good for the device and when the device is ideally placed, showing that the conditions for failure are not readily predictable (De La Cruz et al. 2015).

An SCS delivers an electric current to the nerve that transports the pain signal to the brain. When placed effectively and controlled, the current prevents the pain signal from reaching the brain. In successful cases, the results can be life changing for a patient who has been living with chronic pain or opioid addiction. It can enable that patient to get out of bed, go back to work, and to be socially active. When the treatment fails, however, the results can be catastrophic.

A recent investigation by the International Consortium of Investigative Journalists (ICIJ) provided insight into the high stakes of the risk and damage that can occur. SCSs account for the third-highest number of medical device injury reports to the FDA, exceeded only by insulin pumps and hip implants. In these reports, patients claim to have been shocked or burned or have suffered spinal cord nerve damage. Some suffer from muscle weakness, and others report paraplegia. Since 2008, 80,000 incidents have been reported (Weiss and Mohr 2018).

Heartbreaking interviews were conducted with patients whose SCS implants were unsuccessful. These patients described how their doctors and sometimes even the SCS company representative provided them with expectation that the device would relieve their pain and restore them to a normal life. Instead, many wound up more severely injured, and some must remain in bed or will only walk a few steps for the rest of their lives (Weiss and Mohr 2018).

In its report, the ICIJ indicated that there is a disproportionately high rate of injury among the 60,000 SCSs implanted annually. The investigation revealed that many of the devices were brought to market via the 510(k) process, requiring very little new clinical evidence. Given the severity of these failures, the ICIJ raised concerns about the FDA approval process, drawing into question whether this process is effective and appropriate for a device with such potential risk. SCSs are not the only devices that raise these concerns. The ICIJ further cited reports of more than 1.7 million injuries and nearly 83,000 deaths reported to the FDA over the last 10 years for other medical devices (Weiss and Mohr 2018).

Important Design Knowledge

Anyone who has worked toward bringing a medical device to market in the United States will attest to the significant efforts involved, even for a 510(k) approval. Many will claim that the rigorous amount of activity, testing costs, and duration can be prohibitive and may not justify the development of a product. Furthermore, it has been noted that novelty or innovation can be discouraged because the lack of suitable predicates would prevent the use of this "streamlined process." Without a predicate, a medical device is forced into a Class III classification, which requires much more data and documentation, thus increasing costs and time to market. Nonetheless, the FDA monitors these concerns and reports closely and reacts to them. Going forward, medical devices that have high risks may require significantly more data even if they might qualify as a Class II device for 510(k) clearance.

Medical Device Reclassification

Urogynecologic Mesh

While it has been rare for the FDA to invoke market restrictions, it may become more common in the future. SCSs remain on the U.S. market, under the watchful eye of the FDA. The FDA has demonstrated that it now has the authority to monitor and manage products that, once on the market, appear to exhibit higher risk than expected. The FDA has further demonstrated that it will exercise its authority as necessary. Let's consider the case involving **urogynecologic mesh**.

Pelvic Floor Disorders

Pelvic floor disorders are conditions that affect one in five women (Office on Women's Health 2016). When the muscles that support the pelvic organs become weak, one or more organs might press into the vagina. These conditions include **pelvic organ prolapse** (**POP**), fecal incontinence, and urinary incontinence. Pelvic floor disorders may be a result of childbirth, straining, heavy lifting, or simply aging and the hormonal changes associated with older age. These conditions can cause pain and lower quality of life and interfere with sexual function.

They are particularly devastating because many women are embarrassed to speak about them. Instead, they choose to suffer in silence, without seeking treatment. Some women simply accept that their condition is "normal." A woman may avoid social situations, physical activities, and sex. This may harm a woman's self-image.

Fortunately, for those who are willing to consider treatment options, there are several. In most cases, a doctor would first present possible exercise, diet, or other lifestyle changes. In patients for whom these behavioral modifications are not effective, there are device and surgical options.

Pessary

One form of treatment is a **pessary**, which is a mechanical device that is inserted into the vagina and positioned to uphold the pelvic floor. Ancient forms of pessaries can be traced back to biblical times. Today, there are many types and sizes of pessaries that can be selected for the patient's specific size and condition. Most are made of a strong but pliable inert silicone or other plastic. Top manufacturers include Cooper Surgical, Mentor, and Superior Medical.

A pessary must be selected and fitted by a physician. The physician may then train the patient on insertion and removal of the device. It may be difficult for a woman to insert and remove a pessary on her own; in that case, she must return to her healthcare provider for insertion and removal.

A pessary must be removed, inspected, cleaned, and reinserted at least every three to six months, but it can be removed more frequently at the patient's discretion. All patients should be examined every three to six months for erosion or ulcers (Bordman and Telner 2007).

Other risks include vaginal infection, pain, bleeding, and odor and/or discharge. Proper fitting is crucial to the long-term success of a pessary.

Still, a woman who chooses to use a pessary becomes committed to lifelong frequent visits to her healthcare provider. She may have difficulty inserting and removing the device on her own or may experience pain and bruising during insertion and removal. Women who choose to avoid these experiences many seek a more convenient or less onerous solution.

Surgical Options

There are surgical solutions to pelvic floor disorders. A surgery called **colpocleisis** treats POP by closing the vaginal opening. But this is not practical for many women, although for women who do not plan to have or who no longer have vaginal intercourse, it is an option.

Alternatively, there are surgical procedures aimed at repairing the pelvic floor and restoring the organs to their original position. During surgery, the patient's own body tissue may be used to reconstruct the support to the pelvic floor. That tissue may be reinforced with surgical mesh if warranted. Surgery for prolapse can be done with or without mesh and through either the vagina or abdomen (Office on Women's Health 2016).

Surgical Mesh

Prosthetic material was considered to be a means to close hernia defects as early as the 1890s. However, in those attempts, all failed because of infection, rejection, or recurrence (Baylón et al. 2017). It would take another half century before the right materials were discovered.

Surgical mesh was first used effectively in the 1950s for repairing abdominal hernias. Those first meshes were made of polypropylene. Surgical mesh, as you might expect, is a screenlike material that is implanted to reinforce weak tissue. It can be made of metal, polymer, or biological materials such as porcine tissue, or it can even be made of a composite of various types of materials. It may be placed to provide permanent support, or it may be of absorbable material that dissolves over time to provide temporary support. The mesh serves as a scaffold; the aim is for the host tissue to grow through the mesh, ultimately allowing for the host tissue to rebuild itself to a strength suitable to withstand the tension it undergoes during normal physiologic function (see Chapter 14).

By the 1970s, surgical mesh products that were indicated for abdominal repair found their way into gynecologic surgeries such as those for POP. A surgeon would cut the hernia mesh to the shape needed for pelvic organ support and then place it via an incision in the abdomen. Gradually, mesh manufacturers began producing products designed for gynecologic applications, and near the turn of the twentieth century, the FDA cleared these meshes for gynecologic indications such as stress urinary incontinence (SUI) and POP.

Cleared Based on Abdominal Hernia Repairs

The first transvaginal mesh product was approved by the FDA in 1996. It was a product called the ProteGen Vaginal Sling, manufactured by Boston Scientific for the treatment of SUI. This procedure avoided the need for an incision in the abdomen and instead achieved placement via the vagina. When presented to the FDA, this mesh was compared to mesh used for abdominal hernia repair and similar material used in cardiac surgery.

Boston Scientific successfully argued that its vaginal mesh was substantially equivalent to abdominal meshes already on the market. This allowed for its mesh products to be marketed without any clinical data. The only in vivo evidence was a three-month study on rats. By 1998, Johnson & Johnson had introduced its mesh product, GYNECARE, to the market via a 510(k) process based on its substantial equivalence to ProteGen.

Meanwhile, in June of 1998, the FDA inspected a Boston Scientific plant where ProteGen was manufactured. There had been many reports to the FDA regarding adverse events with the product. In that visit, the FDA discovered that Boston Scientific had been receiving many more complaints than were represented in the reports to the FDA. Still, the inspection report did not indicate that there was a reason for ProteGen to be recalled.

Ultimately, though, after hundreds of reports from women describing vaginal tissue damage, discomfort, and pain, Boston Scientific recalled the product in January 1999. The company later settled more than 700 lawsuits. Despite the recall, GYNECARE remained on the market, and other transvaginal mesh products were introduced without concern for this event. In 2002, vaginal mesh was cleared by the FDA via a 510(k) process for the treatment of POP. Many vaginal mesh devices and kits were subsequently cleared for market in the same manner (Glassman 2020).

Introduction of Transvaginal Mesh (TVM) Kits for POP

By the 1990s, a minimally invasive technique and tool kits for placement of the mesh had been established. The mesh was inserted through a small incision in the vagina; this approach was aimed at reducing recovery time compared with the abdominal approach. Instead of using a series of absorbable sutures to build folds in the connective tissue of the pelvis, trocars were used to fasten a permanent mesh to pelvic support structures overlying the defect (Iyer and Botros 2016).

Originally, the mesh was placed using general surgical instruments. But the delivery method lent itself to some novel designs for surgical tools. New delivery devices were developed that assisted in the placement of a pulley stitch through the supportive connective tissue (Iyer and Botros 2016). Ultimately, surgical mesh for transvaginal POP repair was incorporated into kits that included surgical instruments that were especially designed for transvaginal surgical placement of the mesh.

These new kits, and the surgical technique associated with them, required a different set of skills and training. They called for specialized surgical technique and knowledge and experience in pelvic reconstruction. However, the kits were marketed to all surgeons, many of whom did not have sufficient background.

Complications with Transvaginal Mesh

In October of 2008, the FDA issued a Public Health Notification (PHN) entitled, "Serious Complications Associated with Transvaginal Placement of Surgical Mesh in Repair of Pelvic Organ Prolapse and Stress Urinary Incontinence." In this report, the FDA stated that it received more than 1,000 reports from nine different surgical mesh manufacturers describing complications associated with transvaginal mesh used for the treatment of POP and SUI (Daneshgari 2009).

Complications included erosion, infection, pain, urinary problems, and recurrence. There were also reports of bowel, bladder, and blood vessel perforation during insertion. There were additional claims of patients who suffered vaginal scarring and mesh erosion, which led to a significant decrease in patient quality of life owing to discomfort and pain. Patients with complications had to undergo additional treatment, including additional surgery, mesh removal, drainage of abscesses, and blood transfusions or other intravenous therapy. The FDA indicated that the complications were serious but rare. Physicians were warned to obtain appropriate training and to be vigilant in evaluation and follow-up of their patients.

In July 2011, the FDA upgraded its warning. By this time, it had received nearly 3,000 more complaints. In the upgraded statement, the FDA indicated that these adverse events were no longer rare and that the risk in these transvaginal mesh surgeries is higher than in other surgical options. In that same year, the FDA convened an advisory panel of pelvic surgeons and began an investigation into transvaginal mesh products. At the same time, the FDA also recommended that these products be reclassified until additional testing could be conducted.

In January of 2012, the FDA ordered all transvaginal mesh manufacturers to conduct a long-range three-year comprehensive study to identify the risk and complications associated with transvaginal mesh products. At that time, more than 650 lawsuits had already been filed.

At its peak, more than 100,000 transvaginal mesh procedures were performed annually. Unfortunately, approximately 30 percent of women who had the procedure would need another related surgery because of complications. To date, tens of thousands of lawsuits have been filed. In addition to the injuries, the lawsuits posit that the manufacturers failed to inform the FDA or make public known risks associated with the device. By 2013, settlements in favor of the injured plaintiffs ensued.

FDA Reclassifies Transvaginal Mesh Devices

In a bold move, on January 5, 2016, the FDA issued a final order to reclassify surgical meshes for transvaginal POP repair from Class II to Class III devices "based on the determination that general controls and special controls together are not sufficient to provide reasonable assurance of safety and effectiveness for this device. In addition, in the absence of an established positive benefit-risk profile, FDA has determined that the risks to health associated with the use of surgical mesh for transvaginal POP repair identified previously present a potential unreasonable risk of illness or injury" (FDA 2016).

One year later, the FDA issued another final order regarding the instruments that were developed for use in transvaginal mesh procedures. Those instruments were reclassified from Class I to Class II devices, defining them as "specialized surgical instrumentation for use with urogynecologic surgical mesh" (FDA 2017).

FDA Forces Withdrawal from Market

On April 16, 2019, the FDA issued an order to all manufacturers of surgical mesh intended for transvaginal repair of POP to immediately stop distributing and selling their products. Two manufacturers had remained in the market, Boston Scientific and Coloplast. As required under the Class III reclassification, each has submitted premarket approval applications (PMAs), which were reviewed by a specially convened advisory panel that sought expert opinions on how to evaluate the safety and effectiveness of these products for transvaginal repair of POP.

At this point, the panel was asked to describe the needed support to justify how any benefits might outweigh the risks. The panel recommended two very steep attributes: (1) the effectiveness should be superior to native tissue at 36 months after surgery, and (2) safety outcomes should be comparable with native tissue repair. The FDA stated, "Since the FDA has not received sufficient evidence to assure that the probable benefits of these devices outweigh their probable risks, the agency has concluded that these products do not have reasonable assurance of safety and effectiveness. The companies will have 10 days to submit their plans to withdraw these products from the market." Finally, despite banning the use of this product in patients, the FDA required the companies to continue their follow-up of the subjects who had been enrolled in their 522 studies (FDA 2019). The FDA concluded its order with this statement: "Patient safety is our highest priority, and women must have access to safe medical devices that provide relief from symptoms and better management of their medical conditions. The FDA has committed to taking forceful new actions to enhance device safety and encourage innovations that lead to safer medical devices, so that patients have access to safe and effective medical devices and the information they need to make informed decisions about their care." (FDA 2019).

Important Design Knowledge

There is a large opportunity to improve pessary design. "Although serious side effects are infrequent, insertion and removal of most pessary types still pose a challenge for many patients. Pessary design should continue to improve, making its use a more attractive option" (Jones and Harmanli 2010).

Summary

Despite all the regulations and guidelines that have been introduced over the past half century, we still see medical devices that make it to market that are not safe. These cases—Essure, spinal cord stimulators, and surgical and transvaginal mesh—represent some of the serious issues that have arisen after a medical device had been cleared for market by the FDA via the 510(k) process. Data from longer-range use can dramatically change a device's risk profile.

Different from 50 years ago, however, is the FDA's involvement in addressing postmarket safety issues. The FDA is now able to monitor adverse events with more scrutiny and can assert more authority. Altough the FDA will initially try to encourage the device manufacturer to rectify an issue appropriately, it has demonstrated in recent times that it is willing and able to assert its authority, creating market restrictions, demanding device reclassification, and even forcing market withdrawal. This is a new, "tougher" FDA that is expected to get even tougher. Medical device manufacturers should shift their development efforts to ensure that risk is effectively addressed during the 510(k) application process, and this will include the need for better long-range data.

Even with medical devices that are safely marketed, there are huge needs and opportunities for better solutions, such as for back pain or pelvic floor disorders. These solutions may be instead of, or **adjuvant** to, drug treatments and conservative care. Medical device manufacturers must consider alternative treatments beyond the form of their medical devices when assessing the potential value and impact of their products.

Study Questions

1. How does Essure relate to Chapter 5, where you learned about the Dalkon Shield?
2. What are some of the challenges with using imaging to identify the etiology of back pain?
3. Explain some of the differences in using drugs versus devices versus surgical intervention when dealing with back pain.
4. Is an innovative medical device always the best idea for treating a complex problem? Cite evidence as proof.
5. What is missing from the procedure to get a 510(k) approval that could lead to a product being unsafe?

Thought Questions

1. Long-range data can change the risk profile of a device. Explain how.
2. Research James Sims and his work as the "father of gynecology." What surprises you about this story? Where do you see connections between medical advances discussed in the text?
3. What lead to the opioid crisis becoming as widespread as it is today? How could an effective spinal cord stimulator potentially have mitigated some of the spread?
4. What kind of power currently does the FDA have to handle a device that does not appear to be safe? How is this different from 50 years ago? Reference earlier chapters for a hint.
5. Should postmarket studies be mandatory for all medical devices and drugs? What are the implications and tradeoffs for this requirement?

References

Baylón, Karen, Perla Rodríguez-Camarillo, Alex Elías-Zúñiga, et al. 2017. "Past, Present and Future of Surgical Meshes: A Review." *Membranes* 7(3):1–23. https://doi.org/10.3390/membranes7030047.

Bordman, Risa, and Deanna Telner. 2007. "Pessary Insertion: Choosing Appropriate Patients." *Canadian Family Physician* 53(3):424–425.

Center for Devices and Radiological Health. 2014. "Urogynecologic Surgical Mesh Implants." FDA, Washington, DC. https://doi.org/10.1007/s001250100031.

Centers for Disease Control and Prevention (CDC). 2020. "Understanding the Epidemic." Washington, DC. www.cdc.gov/drugoverdose/epidemic/index.html.

Daneshgari, F. 2009. "Re: FDA Public Health Notification: Serious Complications Associated with Transvaginal Placement of Surgical Mesh in Repair of Pelvic Organ Prolapse and Stress Urinary Incontinence." *European Urology* 55(5):1235–1236. https://doi.org/10.1016/j.eururo.2009.01.055.

Deardorff, Julie. 2013. "Women Report Ccomplications from Essure Birth Control." *Chicago Tribune*, December 22. http://articles.chicagotribune.com/2013-12-22/health/ct-essure-safety-met-20131222_1_essure-conceptus-fallopian-tubes.

De La Cruz, Priscilla, Christopher Fama, Steven Roth, et al. 2015. "Predictors of Spinal Cord Stimulation Success." *Neuromodulation* 18(7):599–602. https://doi.org/10.1111/ner.12325.

Glassman, Jeffery S. 2020. "Transvaginal Mesh Timeline." Law Offices of Jeffery S Glassman. Boston, NA. www.jeffreysglassman.com/transvaginal-mesh-timeline.html.

Healton, C., R. Pack, and S. Galea. 2019. "The Opioid Crisis, Corporate Responsibility, and Lessons from the Tobacco Master Settlement Agreement." *Journal of the American Medical Association* 350(3):293–301.

Herzog, Richard, Daniel R. Elgort, Adam E. Flanders, and Peter J. Moley. 2017. "Variability in Diagnostic Error Rates of 10 MRI Centers Performing Lumbar Spine MRI Examinations on the Same Patient Within a 3-Week Period." *Spine Journal* 17(4):554–561. https://doi.org/10.1016/j.spinee.2016.11.009.

Iyer, Shilpa, and Sylvia M. Botros. 2016. "Transvaginal Mesh: A Historical Review and Update of the Current State of Affairs in the United States." *International Urogynecology Journal* 28(4):527–535. https://doi.org/10.1007/s00192-016-3092-7.

Jones, Keisha A., and Oz Harmanli. 2010. "Pessary Use in Pelvic Organ Prolapse and Urinary Incontinence." *Reviews in Obstetrics and Gynecology* 3(1):3–9. www.ncbi.nlm.nih

.gov/pubmed/20508777%0Ahttp://www.pubmedcentral.nih.gov/articlerender.fcgi?artid =PMC2876320.

Liu, Lindsy. D. Pei, and P. Soto. 2018. "History of the Opioid Epidemic." National Capital Poison Center, Washington, DC. www.poison.org/articles/opioid-epidemic-history-and -prescribing-patterns-182.

Mirza, Sohail K., Richard A. Deyo, Patrick J. Heagerty, et al. 2013. "One-Year Outcomes of Surgical Versus Nonsurgical Treatments for Discogenic Back Pain: A Community-Based Prospective Cohort Study." *Spine Journal* 13(11):1421–1433. https://doi.org/10.1016/j.spinee.2013.05.047.

National Institute of Neurological Disorders and Stroke. 2018. "Low Back Pain Fact Sheet." National Institutes of Health, Bethesda, MD. https://doi.org/https://www.ninds.nih.gov/ Disorders/Patient-Caregiver-Education/Fact-Sheets/Low-Back-Pain-Fact-Sheet.

Office on Women's Health. 2016. "Pelvic Organ Prolapse." US Department of Health and Human Services, Washington, DC. https://doi.org/10.1024/0040-5930/a001037.

U.S. Food and Drug Administration. 2016a. "FDA Takes Additional Action to Better Understand Safety of Essure, Inform Patients of Potential Risks." Washington, DC, February 29. www.fda.gov/news-events/press-announcements/fda-takes-additional-action-better -understand-safety-essure-inform-patients-potential-risks.

———. 2016b. "Labeling for Permanent Hysteroscopically-Placed Tubal Implants Intended for Sterilization." Washington, DC, October 31. https://wayback.archive-it.org/7993/ 20180125103558/https://www.fda.gov/downloads/MedicalDevices/DeviceRegulationand Guidance/GuidanceDocuments/UCM488020.pdf.

———. 2017. "Obstetrical and Gynecological Devices: Reclassification of surgical Instrumentation for Use with Urogynecologic Surgical Mesh. Final Order." *Federal Register*, vol. 82.

———. 2019. "FDA Takes Action To Protect Women's Health, Orders Manufacturers of Surgical Mesh Intended for Transvaginal Repair of Pelvic Organ Prolapse to Stop Selling All Devices." Washington, DC, April 16. www.fda.gov/news-events/press-announcements/ fda-takes-action-protect-womens-health-orders-manufacturers-surgical-mesh-intended- transvaginal.

———. 2020. "Essure Permanent Birth Control." Washington, DC, January 10. www.fda .gov/medical-devices/implants-and-prosthetics/essure-permanent-birth-control.

Weiss, Mitch, and Holbrook Mohr. 2018. "Spinal-Cord Stimulators Help Some Patients, Injure Others." *Associated Press*, November 26. www.apnews.com/86ba45b0a4ad443fad12 14622d13e6cb.

CHAPTER 14

Designing for Biocompatibility and Infection Prevention

Learning Objectives

Learn about surgical mesh.

Distinguish between biocompatible and bioinert.

Recognize how a medical device material can lead to complications such as inflammation, scarring, adhesions, infection, and explantation.

Gain insight into how absorbable materials perform in vivo.

Appreciate how mechanical properties such as porosity, elasticity, and tensile strength affect a device's biocompatibility.

Learn how common allergies lead to device failure (e.g., nickel allergy).

Learn about generally accepted practices for cleaning, disinfecting, and sterilizing and how these practices are distinguished by the Spaulding classification.

Realize the impact cleaning, disinfection, and sterilization processes may have on medical devices and how these considerations influence design decisions.

New Terms

bioinert

biocompatible

adhesions

catgut

polyglycolic acid (PGA)

polylactic acid (PLA)

polydioxanone (PDO)

caprolactone

Wolf's law

stress shielding

Nitinol

Naval Ordinance Laboratory (NOL)

Spaulding classification

flash sterilization

brachytherapy

interstitial brachytherapy

Biocompatibility

The objective for any implanted device is to completely restore tissue, organ, or structural function, although, in reality, this is very difficult to achieve. Device material is very significant to the overall achievement of this goal. Often there are tradeoffs between a material that is best for the mechanical performance of a device versus its biologic performance. No matter how good the mechanical design of a medical device may be, if the device's material is not biologically compatible, the device's effectiveness and safety may be severely compromised.

In Chapter 13, we learned about some of the challenges associated with devices made of surgical mesh. Still, this material offers some useful advantages because of its open-weave structure. For example, when surgical mesh is used to repair a hernia, optimally it would allow for the host tissue to weave through the pores of the mesh, ultimately growing into new abdominal muscle and connective tissue that is just as strong as the original tissue was prior to injury or damage. If the mesh were made of degradable material, ideally, that material would gradually be absorbed or dissolve without incident while the host tissue heals to restore normal performance. Alternatively, a nondegradable mesh would integrate harmoniously with the host's living tissue without any negative reaction.

More than 700,000 abdominal hernias are repaired with surgical mesh each year in the United States. Despite the huge number of successful abdominal hernia repairs with mesh, the rate of adverse events is quite high. Hernia recurrence rates may exceed 40 percent, and failure rates may exceed 10 percent. The most common adverse events following hernia repair are pain, infection, hernia recurrence, adhesion, and bowel obstruction. Mesh migration and mesh shrinkage also may occur (U.S. Food and Drug Administration [FDA] 2018).

Many mesh products have been recalled. The complications described have been attributed largely to mesh products that are no longer on the market. Note that in an analysis performed by the Food and Drug Administration (FDA) involving medical adverse event reports, it was revealed that the main cause of bowel perforation and obstruction was recalled mesh products. Thankfully, today many of the meshes responsible for these adverse events are no longer being used.

Compared with nonmesh surgeries, surgical mesh repairs have reduced the rate of hernia recurrence significantly, and mesh is now used in 80 to 90 percent of all abdominal hernia repairs. More than 70 types of meshes are currently available commercially. Still the rates of adverse effects such as infection, adhesion, and bowel obstruction remain higher than desired. Most of these effects are related to the chemical and structural nature of the mesh itself (Baylón et al. 2017). Medical device engineers and scientists continue to strive to identify materials and other characteristics that could improve outcomes.

Bioinert versus Biocompatible

When a foreign material is introduced into the body, it triggers a healing response. It will elicit one of three types of reactions: (1) destruction, (2) inclusion or tolerance, or (3) rejection or removal (Baylón et al. 2017). If the host tissue does not react at all to an implanted material, the material is considered to be **bioinert**. In reality, however, there are few, if any, truely bioinert materials. Note that bioinert materials are not necessarily ideal because they might not integrate, or "stick," at all to host tissue involved. The aim of most implantable materials is to obtain some integration into the host tissue but to avoid negative host reactions such as inflammation or infection. This would represent the most desired reaction—inclusion or tolerance—and therefore be considered an ideal and **biocompatible** material.

Inflammation

Any time tissue is injured, even as a result of surgery, there will be an acute inflammatory response. Ideally, this response will lead to normal healing of the wound. When healing doesn't occur, the inflammatory process may progress into chronic inflammation. When chronic inflammation is related to an implanted foreign material or medical device, the host will undergo an immune response and will deliver cells to combat the foreign body.

Scar

Ultimately, some of the special cells involved in the immune response will produce scar tissue. This scar tissue does not have the same desired properties of normal host tissue. Scar tissue is generally stiffer and less organized, leading to poor mechanical integrity of the tissue and the implanted system. In the example of surgical mesh, this lack of strength can lead to recurrence of a hernia.

Infection

Meanwhile, the chronic inflammation may manifest as an infection. As we discussed in Chapter 12, the presence of a substrate combined with compromised tissue may allow for bacterial growth that can lead to infection. Infection delays or prevents the wound-healing process.

Adhesions and Obstructions

After abdominal surgery, scar tissue may develop in undesired locations. Scar tissue could cause organs and visceral tissue to stick where they shouldn't, creating **adhesions**. Scar tissue can cause blockages, such as obstructions of the intestines, or it can cause abnormal connections between organs, blood vessels, or intestines. This can lead to pain, fluid buildup, bleeding, or perforations.

Explantation

With any severe adverse reaction, infection, or recurrence, the implanted material must be removed. Removing surgical mesh can be difficult and risky. The mesh may be integrated with organs and abdominal tissue from which it must be cut or extracted. Many engineers and scientists are working to develop materials that are more biocompatible. The research field is rich with new ideas regarding coatings, textures, patterns, chemical compositions, and material constructs such as fibers and gels. Successful translation of these ideas into a surgical mesh with optimal performance has yet to be developed.

Absorbable Materials

Earlier we mentioned that a surgical mesh may be made of an absorbable material. The intention of an absorbable material is to provide temporary mechanical support and to provide a substrate onto which the host tissue may grow. Over time, the host tissue would grow strong enough to perform as needed, and the absorbable material would gradually degrade and disappear.

In actuality, the absorption of a foreign material is part of the inflammatory response. The material is exposed to conditions in the body—physical changes and cellular activity—that help to break down and remove the material. An ideal absorbable material does not cause a high level of chronic inflammation, and the body tolerates it without an adverse reaction.

Absorbable Sutures

You have probably had or heard of absorbable sutures. In fact, the first absorbable sutures were made of catgut. **Catgut** is made of animal intestines, usually sheep (never from cats). It has been used as string for tennis racquets and musical instruments and likely derives its name from the early string instruments on which it was used.

Catgut is still used in sutures today, but now absorbable sutures may also be made from the following polymers or blends thereof: **polyglycolic acid (PGA)**, **polylactic acid (PLA)**, **polydioxanone (PDO)**, or **caprolactone**. Absorbable sutures can, on occasion, cause inflammation leading to rejection by the body. Nonabsorbable sutures are often preferred because they are generally believed to result in less scarring because of a lesser immune response compared with absorbable sutures. The actual evidence for this argument is debatable. Still, based on this common belief, many physicians might use nonabsorbable sutures where cosmetic outcome is important.

Important Design Knowledge

In practice, it is extremely difficult to predict the degradation rate and characteristics of absorbable materials in humans without clinical experience. The environment in vivo is an extremely complex and dynamic one, not only with cellular changes but also with changing physical conditions including pH, temperature, moisture, and shear forces, to name a few. Some materials remain far longer than expected, leading to concerns of tissue irritation and incomplete wound healing. Other materials may break down differently than planned, compromising the mechanical performance of the device and leading to the opportunity for reinjury.

Furthermore, conditions may vary widely in one wound depending on the timing or degree of inflammation and can be quite different from one wound site or individual to another. While animal studies are helpful, results can still be substantially different in humans. It is generally quite difficult to accurately replicate all these conditions in bench studies, and it becomes easy to see how problems with absorbable devices that have been cleared for market without clinical data, such as those described with surgical mesh, become evident only after the device is on the market and in patients. To date, there are few models to provide accurate predictors of absorption performance; this purpose is best served with clinical evidence.

Porosity

When selecting a medical device material, it is not enough to simply consider its chemical composition. We have already discussed surface texture characteristics in Chapter 12. The porosity of a material plays a key role in the successful integration of tissue. Cell proliferation is very dependent on pore size. An optimal pore size will allow fibroblasts to bridge gaps in damaged tissue. Keep in mind that bacteria also like to grow and colonize in very similar conditions to those that are optimal for tissue growth. Therefore, a further balance must be struck between an optimal pore size and other measures that will prevent bacterial growth.

Elasticity and Tensile Strength

Earlier we reviewed how deterioration of the tensile strength of an abdominal mesh could potentially lead to hernia recurrence or poor function. Therefore, the materials selected for the device must permit the device to exhibit the mechanical properties necessary to withstand the stresses placed on the abdominal wall; it should be strong enough to withstand normal intraabdominal pressures. Of importance, however, is to note that you should not overdesign the strength of a device. The material should allow for normal tensile strength and distension as well.

Stress Shielding

In some medical devices, there is another reason to avoid selecting a material that is too rigid. You may have already learned about **Wolf's law**, which describes how bone tissue grows in response to stress. If there is not enough load on the bone, it will resorb. But if an orthopedic implant is too rigid, it can prevent the bone tissue from seeing the stress it needs to trigger growth. Instead, **stress shielding** occurs, and the result is that the bone tissue does not grow, and even bone resorption can take place. The best material will meet all the criteria for biocompatibility and will have an elastic modulus that is closely matched to the host bone.

Allergies Cause Device Failure

Sometimes a material that may be most ideal for a device could cause an allergic reaction in some people, leading to device failure. You might have heard of latex allergies. Latex is derived from natural rubber and has been used in many medical devices, including protective gloves. Ironically, latex allergies were not very common until the 1980s, when many healthcare workers began wearing latex gloves to control infection. The widespread use of latex gloves led to the common occurrence of latex allergy; now we know that around 6 percent of the population is allergic to latex.

An allergy is a heightened immune response to an invader or allergen. This response may be invoked when exposure exceeds threshold concentrations and duration. There is a range of normal threshold values for humans, but the threshold for some people is outside that range. Some people will have an allergy to what is typically harmless to most other people; they just have a genetic makeup with a lower threshold that makes them sensitive to that invader. Others can become more sensitive to an allergen from exposure to excessive levels or even to lower-level exposure over a long period of time.

Metal Allergies

Metal allergies have been associated with the failure of devices such as hip and knee implants. Some of the most common metals in these prostheses are chromium, nickel, cobalt, titanium, and molybdenum. Often a prosthetic joint will contain more than one of these metals. Around 2 percent of the population is allergic to cobalt or chromium. The most common metal allergy is to nickel.

Nickel

Between 10 and 20 percent of the population is allergic to nickel; this number is growing as device wearables such as smartwatches gain popularity. Currently, more women than men have nickel allergies; this disparity is attributed to the increased exposure that most women have from wearing jewelry.

Nitinol

Perhaps you are familiar with **Nitinol**. This is an alloy composed of nickel and titanium. In the early 1960s, the **Naval Ordinance Laboratory** (**NOL**) was trying to develop a metal for heat and corrosion resistance. Nickel is commonly added to alloys such as stainless steels to reduce rust, for example. It made sense to incorporate nickel with titanium to enhance corrosion resistance. Coincidently, it was discovered that this new alloy, *Ni-Ti-Nol* (named for the laboratory where it was discovered) exhibited unique and useful shape memory and super elastic properties.

Guidewire

Consider the usefulness of a material such as Nitinol. During an interventional cardiology procedure, a surgeon must navigate a balloon, stent, or other device through tortuous vasculature. From an incision in the femoral artery, the surgeon must steer through approximately three to four linear feet of blood vessels in order to reach the heart. The first step in these surgeries often is to insert a guidewire. Of course, the wire must not puncture the vessel during navigation and cannot get stuck. Early guidewires were made of stainless steel and often developed kinks in use. If a kink developed, the wire had to be carefully removed, and the insertion procedure had to restart from the beginning. With greater elasticity, a guidewire made of Nitinol could flex and rotate more smoothly, with less risk of kinks, getting stuck, or puncturing a vessel.

The same elastic properties make Nitinol an ideal material for devices delivered to the body through the vasculature. Snares, retractors, and baskets can be temporarily deformed, or squeezed down, to fit into a blood vessel until they reach the site where they must open to remove foreign bodies or blood clots. At the site, they spring back to their original shape to capture the obstruction.

Stents and Restenosis

Similarly, a stent can be delivered for placement, and on reaching the constricted area of a vessel, the stent can be deployed by springing back to shape; in this way, it serves as a scaffold to hold the vessel walls open. While the data are not completely clear, there is concern that such exposure to nickel can trigger an allergic response in the endothelium of sensitive individuals, leading to potential restenosis or other undesired adverse effects (Koster, Vieluf, and Sommerauer 2001).

Important Design Knowledge

When selecting materials for a device, you must consider allergies and other negative reactions. In the preceding example, the nickle-titanium alloy was good because it was flexible and springy but bad because a small subpopulation of patients is allergic to nickel. Nitinol is used in medical devices today, although it is more widely accepted in tools and instruments—such as guidewires—that are not permanently implanted. No matter how small the population, doctors may not want to take any potential risk with their patients. Even if the potential risk is not fully substantiated in the literature, a surgeon may be deterred from using a product with a material that might pose risk of a bad reaction. This sentiment may limit clinical adoption of such products, constraining their marketability.

Cleaning, Disinfection, and Sterilization

We have seen that sometimes, even when a medical device is properly sterilized, its mere presence can allow for colonization by indigenous bacteria and other microbes. We have reviewed how an infection caused by a medical device can lead to sepsis, explantation, revision, and other life-threatening complications. For these reasons, the ability to clean, disinfect, or sterilize a medical device and to maintain a clean or sterile environment plays a significant role in medical device design.

Note that there is a difference between sterilization, disinfection, and cleaning. A sterilization process must kill all forms of microbes, many of which can be very resilient. Disinfection eliminates many or all pathogenic organisms but not spores or inanimate objects. Cleaning reduces the bioburden and removes foreign material. Cleaning should be done before the disinfection or sterilization to minimize interference with these processes (Centers for Disease Control and Prevention [CDC] 2013).

Spaulding Classification

It is not necessary to sterilize all patient care items. In 1968, a scientist named Earle Spaulding, who had extensively studied these processes for medical devices, wrote a hierarchy of requirements that were dependent on the type of contact the device would have with the body (Table 14.1). This scheme is referred to as the **Spaulding classification** and is the basis for current CDC and FDA guidelines.

TABLE 14.1   Spaulding's Classification Established Guidelines for Cleaning, Disinfecting, and Sterilizing Medical Devices

Body Contact	Disinfection Requirements	Medical Device Class
Intact skin	Low level	Non-critical
Mucous membranes	High level	Semi-critical
Sterile body cavity	Sterilization	Critical

Items associated with a high risk of infection are critical items; as we have seen, implants and surgical instruments are good examples of critical items. An example of a semicritical item is one that comes in contact with skin that is not intact or with mucous membranes, for example, lungs. A respiratory therapy device would fall into this class. A device that comes in contact with intact skin falls into the noncritical class because skin is an effective barrier to most microorganisms. These items, such as bedrails and blood pressure cuffs, can be cleaned (Rutala and Weber 2016). However, care must be taken to avoid secondary transmission from a person who has touched these items to a patient with a more critical risk from exposure to a person who has touched those items. This is why it is so important for healthcare workers—and for you—to wash your hands!

Designing for Challenges to Cleaning, Disinfection, and Sterilization Processes

Protect the Device from Disinfection and Sterilization Processes

Disinfection and sterilization processes, by their very nature, can be harsh. After all, the processes need to effectively kill many or all microbial agents. A medical device design must always consider the need for disinfection or sterilization. There are many different methods for each. A process must be selected so that the device is unharmed, and conversely, a device must be designed to withstand the process. For multiple-use devices, the design must ensure that the device can withstand many process cycles.

Common Sterilization and Disinfection Processes

It is important to understand the environment in which these processes occur so that a device may be designed to withstand the conditions it will see during sterilization or disinfection. These treatments may involve heat, steam, chemicals in liquid and gaseous form, and radiation. They may cause subsequent reactions or leave behind harmful by-products that must be managed. For example, treatment with radiation can yield ozone, which must be properly contained.

Steam Sterilization.   Steam sterilization is a predominant method of sterilization, especially in the clinic. It is dependable: nontoxic, inexpensive, microbicidal, sporicidal, and rapidly heats and penetrates fabrics (CDC 2013). While this may be ideal for metal hardware and for bed sheets, steam could be damaging to many medical devices that include plastic in their makeup. Steam treatment introduces high temperatures and moisture, and it is possible that a medical device might not tolerate exposure to either or both of these elements. Since the introduction of polymer materials in medical devices in the 1950s, alternative low-temperature methods of sterilization have been developed.

Central Sterile.   Steam remains the most common sterilization method in most clinics. Larger hospitals or institutions usually have a central location to perform the sterilization processes. They have very strict procedures starting from receipt of contaminated devices to preparation, treatment, and then wrapping, storing, and transporting. If you are involved with surgical devices, you may already be familiar with the large trays full of tools and instruments that are used during an implant surgery.

The device manufacturer is required to provide instructions for preparing and sterilizing its instruments, which must account for water drainage and placement of heavy items relative to more delicate items. Overall weight of a tray/instrument system must also be considered. One of the most frustrating issues is for a tray to be delivered to the operating room under the assumption that it is ready for use, but once unwrapped and opened, it is wet. A wet tray is not sterile despite having gone through the sterilization cycle, and it cannot be used. A heavy metal mass in the tray is often the cause of a wet pack. Sometimes this is due to a few misplaced or added instruments in the tray. The design of a medical instrument system must account for the reality of the use cases, not just ideal cases.

Flash Sterilization.   Another sterilization reality is the use of **flash sterilization**. This is a modification of conventional steam sterilization. It is sometimes used when there is insufficient time to sterilize an item by the preferred method. Flash sterilizers are small units that are often positioned in or near operating rooms and may be used when there is an urgent need. For example, think about the conditions under which surgical instruments are used. Surgeons and assistants wear multiple layers of gloves, and their hands and instruments are exposed to blood and other bodily fluids. This sometimes creates slippery or sticky handling. If an instrument has been used and on being returned to the sterile field is inadvertently dropped outside the sterile field, it might be sent for flash sterilization so that it could be used again later during the same surgery.

Flash sterilizers are not to be used merely for convenience. While they introduce steam, control over the process is different from the manufacturer's instructions. We have just reviewed how exposure duration, temperature, and even geometry may affect the efficacy of a sterilization process. For these reasons, there is a higher risk of potential infection. Flash sterilization should never be used on implantable devices. Nonetheless, the possible use of a flash sterilizer should be in the mind of a designer of instruments and other nonimplantable medical devices.

Common Lower-Temperature Processes

Common sterilization processes include ethylene oxide gas, liquid immersion in a chemical sterilant such as glutaraldehyde, and e-beam or gamma radiation. These are just a few examples. There are many different methods, and each has advantages and disadvantages. You should review each thoroughly to be sure that the device design and conditions of use are suitable for the method you select. Similarly, common disinfection processes include pasteurization and, once again, liquid immersion in chemical sterilants such as glutaraldehyde. But for disinfection, the exposure is to lower concentrations of the chemical and for a shorter duration than sterilization.

Time and Concentration

Take note of these differentiators: time and concentration. These are common factors that affect the efficacy of sterilization and disinfection processes. These are also factors that affect the medical device design from material selection and even all the way through to the end of the manufacturing process. A device may be able to withstand exposure to a moderate dose of radiation for a short duration, but it may be damaged by higher or longer exposures. Conversely, concentrations that are too low or exposure durations that are too short could allow for some of the bioburden to survive treatment, rendering the treatment useless.

Sterilization in Package

Now consider that you are designing a medical device that must be sterilized after it is manufactured and before it is shipped. Products like this are often sterilized in the package. This is very common for disposable medical devices or device components. To be disposable, they need to be made of inexpensive materials—usually polymers. As mentioned earlier, it is most likely that the polymer would not withstand the heat that steam introduces, and the package would not tolerate the moisture that steam introduces, so a lower-temperature solution is needed.

If radiation were used, the device and package would need to be considered as a system to ensure effective results. If the package density is not optimal, the efficiency and dose uniformity would suffer (Devitt 2019).

Imagine that you have been asked to update the design of a medical device that has already been commercialized. This is a very common job for a medical device designer. Manufacturers often need to create a next-generation device in order to address market feedback or acquire new markets. Let's say that you need to make a new, extralarge version of a marketed device to serve an obese population segment. You cannot assume that the same package and sterilization method will work. In fact, small changes, such as geometry or material type, density, and weight, could all affect the effectiveness of sterilization (Devitt 2019).

Location of Microorganism.   Microorganisms that reside in crevices, joints, channels, or lumens can be difficult to reach during any of the aforementioned antimicrobial processes. For this reason, devices and equipment with flat surfaces are preferred. Device manufacturers are encouraged to produce equipment engineered for ease of cleaning and disinfection (CDC 2013).

If you were asked to make a catheter for children based on an adult-sized catheter that is already on the market, you would need to be especially concerned for changes to the lumen. Let's assume that the adult device is sterilized with ethylene oxide gas. If you need to reduce the diameter of the lumen as you make the device smaller, you might find that you have to significantly increase the device exposure time. This could affect the effectiveness of the sterilization process given the device materials and packaging (Devitt 2019).

Microorganism resistance does not vary only by exposure time and duration. Resistance also changes depending on microorganism type and concentration. We have previously discussed how the formation of a biofilm can significantly change the resistance of a microorganism. Other factors that may affect a microorganism's resistance include temperature, pH, relative humidity, and water hardness. Organic or inorganic matter can act as a barrier to the surface needing treatment or can react with a germicide, reducing its potency (CDC 2013).

Important Design Knowledge

A sterilization or disinfection process must be selected to treat the most resistant microorganism, and the device must be designed to tolerate the most robust process anticipated. Be sure that all specific factors related to device design, sterilization or disinfection processes, and microorganism performance are considered when you design a device for manufacture and market. Note that the sterilization or disinfection processes must be reevaluated even for a small change to a device design.

Breast Template Loses Transparent Characteristics after Multiple Sterilization Cycles

Sometimes the damage to a device is not evident until after several sterilization cycles. On August 9, 2013, Varian Medical Systems announced a voluntary recall of its VariSource Advanced Breast Template System. The recall was initiated after Varian Medical "became aware the Lexan templates of the Breast Bridge Template System could become bent and turn opaque after steam sterilization" (U.S. Food and Drug Administration [FDA] 2013).

The system was used in **brachytherapy** procedures. Brachytherapy is a type of radiation treatment for cancer. A radioactive material is inserted into the body near the site where the cancer cells have been identified. Compared with externally projected radiation, this type of radiation is more specifically targeted. This allows for the application of higher doses of radiation to the cancerous tissue while providing less radiation exposure to healthy tissue.

Interstitial brachytherapy is often used in the treatment of breast cancer. A radioactive device is placed within the tissue; the radioactive material is delivered via a needle or tube or similar applicator. The amount of radioactive material is very small; imagine something about the size of a grain of rice. Conventional placement of the catheters or needles is completed freehand. By using a template, the geometric placement of the catheters/needles can be exact, optimizing the physician's ability to target and control the delivery of radiation to the tumor site. The template holds the breast physically in place and is used as a navigational aid, along with diagnostic imaging such as a computed tomographic scan or ultrasound imaging to place the radiation material near the tumor (Varian Medical Systems 2018).

Important Design Knowledge

The VariSource template was made of clear Lexan (a polycarbonate) to allow for the ability to see through the device, enabling visual guidance to its targets and markings. But from multiple sterilization cycles, the Lexan degraded, turning opaque and warping or bending. A guide template is hardly useful in this condition. If someone tried to use this system after it was damaged, the radiation material might not be placed accurately. Varian has corrected the issue and now offers an updated product. Meanwhile, this is a good example to demonstrate why multiple sterilization cycles must be considered during device material selection.

Summary

No matter how well thought out a device design might be mechanically, it can lead to inflammation, scaring, adhesions, infection, explantation, and even death if it is not biocompatible. Medical device design considerations therefore must include material selection and mechanical properties such as porosity, surface characteristics, and elasticity. Still, different people may react differently to certain materials or designs, and a commercial medical device must be safe for all users, including those who may have allergies or otherwise have an inflammatory response—especially in response to implanted, long-duration-contact devices.

Furthermore, cleaning, disinfection, and sterilization practices should be considered when designing a medical device. It is critically important that device manufacturing and use cases be considered in the device design. A device should not be able to harbor or provide "hiding places" for microorganisms. Nor should it be subject to damage during sterilization or cleaning processes that can sometimes include very harsh chemical, temperature, moisture, and radiation treatments. Practical use considerations after the device leaves the manufacturer must also be considered, such as the need to flash sterilize the device in the operating room.

Study Questions

1. What is the different between a substance being biocompatible and being bioinert?
2. Describe the differences between cleaning, disinfecting, and sterilizing an item. Reference the Spaulding classification.
3. How could an allergy lead to a device failure? Cite an example from the text.
4. What is catgut?
5. What can happen if proper antimicrobial procedures are not considered when designing the packaging for a medical device?

Thought Questions

1. Chapter 13 discussed inserting surgical mesh transvaginally. What made this case different from the hernia mesh discussed in this chapter?
2. Describe the process of an ideal absorbable material in vivo. What potential challenges can occur?
3. What happens to tissue after a foreign device is implanted if it promotes an inflammation event in the body? Walk through the body's reaction from insertion onward.
4. How can the mechanical properties of a device affect biocompatibility? Think of the example of porosity, elasticity, and tensile strength.
5. Your organization is charged with designing a balloon catheter for a stent. What are the pros and cons of various antimicrobial techniques? What kind of failure(s) would you want to avoid?

References

Baylón, Karen, Perla Rodríguez-Camarillo, Alex Elías-Zúñiga, et al. 2017. "Past, Present, and Future of Surgical Meshes: A Review." *Membranes* 7(3):1–23. https://doi.org/10.3390/membranes7030047.

Devitt, Shaun. 2019. "7 Elements to Consider When Designing Medical Products for Sterilization." Med Device Online, January 30. www.meddeviceonline.com/doc/elements-to-consider-when-designing-medical-products-for-sterilization-0001.

Koster, R., D. Vieluf, and M. Sommerauer. 2001. "Nickel and Molybdenum Contact Allergies in Patients with Coronary In-Stent Restenosis." *ACC Current Journal Review* 10(3):62. https://doi.org/10.1016/s1062-1458(01)00267-7.

Rutala, William A., and David J. Weber. 2016. "Disinfection and Sterilization in Health Care Facilities: An Overview and Current Issues." *Infectious Disease Clinics of North America* 30(3):609–637. https://doi.org/10.1016/j.idc.2016.04.002.

U.S. Centers for Disease Control and Prevention (CDC). 2013. "Guideline for Disinfection and Sterilization in Healthcare Facilities, 2008: Miscellaneous Inactivating Agents." Atlanta, GA. May, pp. 9–13. https://doi.org/1.

U.S. Food and Drug Administration (FDA). 2013. "Class II Device Recall: Breast Template System." Washington, DC, October 24. www.accessdata.fda.gov/scripts/cdrh/cfdocs/cfRES/res.cfm?id=121521.

———. 2018. "Hernia Surgical Mesh Implants." Washington, DC, February 4. www.fda.gov/medical-devices/implants-and-prosthetics/hernia-surgical-mesh-implants.

Varian Medical Systems. 2018. "Varian BrachyTherapy Applicators and Accessories." PaloAlto, CA. www.varian.com.

CHAPTER 15

Designing for the Use Case

Learning Objectives

Recognize the unique constraints and challenges in clinical use of a device.

Understand how system compatibility and digital image file storage and transfer capacities limit device communication or functionality.

Appreciate how complex device maintenance procedures and lack of standardization throughout an organization can increase the risk of human error.

Learn about the unique characteristics in a clinic that limit accessibility to WiFi, shared and support equipment, and power supplies.

Recognize why patients may be nonadherent and understand what design changes can be made to improve patient adherence.

New Terms

discectomy

pedicle screws

allograft

intensive care unit (ICU)

electromagnetic interference (EMI)

focused assessment with sonography for trauma (FAST) examination

golden hour

nonadherent

obstructive sleep apnea (OSA)

continuous positive airway pressure (CPAP)

endogenous

exogenous

needle-free injection technology (NFIT)

Equipment Management in the Clinic

In Chapter 14, we discussed how sterilization can damage a medical device. If you have worked in a clinic, you might be familiar with some of the other unique challenges regarding the use of medical devices. The need to be able to find and use equipment quickly or even maintain functionality during patient transport can add extremely challenging design criteria to what might otherwise be simple. The critical nature of the clinical environment can test even the most robust devices. If several pieces of equipment are to be used together, the designer must consider the need for communication between pieces of equipment as well as the need to avoid conflict among them.

System Compatibilities

One of the biggest challenges that has arisen from the application of technology in medicine is system compatibility. Years ago, there were only a few sophisticated technologies available for patient care, and most stood alone/independently. Equipment designers had the liberty to develop what they felt was optimal for the performance of that individual piece of equipment without considering the need to integrate that piece of equipment with anything else.

Let's consider this scenario. You are an avid skier, living in the 1970s in the United States. You just had a very hard fall. Your wrist hurts, but you are not sure how badly you are really injured. You certainly don't need an ambulance, but later in the day you decide that you should probably have it checked out. So you go to the emergency room, where the doctor takes an x-ray.

Your x-ray image is captured on a special film that is coated with an emulsion containing silver and other chemicals that are sensitive to x-ray energy. To render an image after exposure, the film must be processed using very specific wet chemistry in a darkroom. When the doctor in the emergency department comes to speak to you, she tells you that she has looked at your x-ray and may have some good news. From what she can see, it does not appear as though your wrist is broken. But the film is still wet, and to be certain, the doctor has to wait for it to dry and for a radiologist to come in and read it tomorrow morning.

The next day, you receive a phone call from the hospital. The doctor explains that the radiologist discovered a hairline fracture in your radius. The doctor suggests that you can come to the hospital to pick up your x-rays so that you can bring them to your orthopedist. Or if you are unable to pick up your films, the hospital will send them to your orthopedist, but that might take a few days.

Despite how antiquated this may seem today, it is an example of what a top standard of care was just 50 years ago. In fact, in other parts of the world, this still represents the highest quality care imaginable, although by the early 1980s x-ray imaging began taking a digital form, creating ripe opportunities to improve standards of care.

Evolution of Digital Imaging

In fact, if your ski accident occurred today, instead of going to the hospital emergency room, you might consider going to an urgent care center or even directly to your doctor's office. Many of these facilities have suitable x-ray and other imaging capabilities right in their offices. The x-ray image would be processed instantly on a computer; there is no need to wait for wet-film processing. Nor is there a need to wait for latent images to evolve as the film dries; the film can be read immediately. Even though there would not be a radiologist on site, the image would be transmitted to a radiologist in any other part of the world who could read it and prepare a report right away.

In a very short period of time, while you wait, the doctor who is treating you would have gotten a definitive image and report about your fractured radius and could immediately apply a cast or other nonsurgical stabilization. Even if you needed to go see a different doctor for treatment, you would not have to physically carry your film with you. The images could be sent simply via the internet.

But wait. It hasn't always been this simple. Initially, the new digital x-ray equipment was developed without consideration for the need to communicate with other systems. The visualization software was customized for that individual device. The image might be read on that device or on like devices elsewhere. Otherwise the image had to be printed in order to be read.

At the time when hospital and medical institutions were investing in the first generation of digital imaging equipment, the need to communicate with other equipment was minimal. These machines were very expensive pieces of equipment designated to serve the entire institute—all of which had access to the same reader technology. Patient records were in physical, paper files, and a printed image was stored with the patient's file. If needed, the patient's images could be transported in the same manner as film images had always been shared, and that seemed to be sufficient. Focus was on enhancing image quality but not on communication of those images. Even though digital technology was available in some areas, it was not yet a part of everyday life, so this shortcoming was acceptable.

Just 10 years later, however, this inadequacy could no longer be overlooked. Digital communication had become mainstream, and frustrations grew as compatibility issues among visualization systems became evident. Keep in mind that the data requirements for these files were quite large, and computer storage and transmission capacities were much smaller than they are today. In fact, at that time, most files were too large to transmit via the internet.

Still, it was far more convenient to carry a small floppy disk and later a small CD containing medical images rather than carrying a bunch of films. It was also much easier to place and store that disk or CD in a patient's file folder than to store films.

The problem, however, was that the software needed to read the images was not the same from one piece of imaging equipment to the next. In fact, certain manufacturers might not have standardized software between different models of their x-ray machines. Furthermore, the hardware and storage requirements for that software usually exceeded the capabilities of everyday office computers. So, if doctors wanted to buy and install the manufacturer's software on their computers, they couldn't, or they would have to purchase a supercomputer to do so!

As digital x-ray technology matured, manufacturers began seeking to standardize their visualization software among their product lines and also across the industry. Meanwhile, advances in digital technology created capacity for more data storage and transmission. Less than 50 years ago, you would have taken several hours or even days to coordinate and complete your acute treatment. Today, your doctor can read simple x-ray images in real time on a smartphone! Your doctor can even speak to you face to face from a remote location. The doctor might even be at the ski slope himself or herself while speaking to you.

Important Design Knowledge

System compatibility must always be considered when designing medical devices. There are still many unresolved system issues in hospitals and clinics. Many of these issues are due to hospitals having different models, vendors, and vintages of equipment that are not standardized.

Patient Monitors and Lack of Standardization

One example of lack of standardization is patient monitors. A hospital department may purchase a large quantity of monitors as a group but will not buy enough to replace every monitor in the hospital—that would be too expensive. As time goes on, departments change and move, and those patient monitors will be moved, too. Without standards, however, the new monitors may perform differently from earlier monitors when integrated into the hospital's various systems. Also, caregivers and equipment maintenance engineers need to be trained differently for each type of monitor. This creates vulnerability, where improper connection, use, or maintenance can lead to a risk of critical errors (Loughlin and Williams 2011).

Dangers of Incompatibility

In 2001, a six-year-old boy in New York was killed in an magnetic resonance imaging (MRI) chamber by a metal oxygen tank. Ferromagnetic metals are banned from MRI rooms. The tank was inadvertently carried into the MRI room by a staffer. The MRI machine's 10-ton electromagnet caused the oxygen tank to fly into the chamber. In this case, the oxygen tank was not compatible with the MRI machine. The staffer either was not trained or had a lapse in judgment, and there were not enough safeguards in place to protect against this catastrophe. The family settled a lawsuit for $2.9 million, and the hospital was fined $22,000 by the state health department (Mangan 2001).

Important Design Knowledge

While catastrophic accidents like this are not common, mishaps do occur more often than they should. As equipment becomes more enriched with technology, incompatibilities may not be so obvious, and the risk of human error is heightened. Maybe bringing a metal oxygen tank into an MRI room is an obvious mistake to most of us, but if you think about the potentially stressful circumstances that may occur, especially during an emergency, it is easy to make mistakes. Imagine if a patient went into distress during an MRI scan and there was an immediate need for another doctor and for patient transport. Everything brought in the room must be nonferrous, from the wheelchair or stretcher right on down to the doctor's stethoscope. Medical equipment designers must consider all users and stakeholders as well as the use environment to prevent misuse.

Device Maintenance

When you think about designing a medical device, you need to consider a complex mix of medical needs, technologies, and regulatory requirements. Yet, despite the very sophisticated design challenges, everyday use of these devices can be curtailed by seemingly simple issues. Clinical and biomedical engineers cite broken connectors that are hard to replace and batteries that often run out of charge as some of the most common reasons why a device cannot be used (Loughlin and Williams 2011).

At the other end of the spectrum, complex equipment often requires a lot of maintenance. Sometimes preventive maintenance can take several hours to perform or must be done very frequently. Equipment designed to deliver precision performance such as robots and imaging systems can be inaccurate if they are not calibrated correctly. Brachytherapy was discussed in Chapter 14. Imagine the consequences of improperly administered radiation therapy. If the delivery machines and their software are not adequately maintained to perform according to specifications, normal tissue can be damaged or tumor treatment could be ineffective. According to hospital biomedical engineers, keeping up with maintenance on such systems is a big challenge (Loughlin and Williams 2011).

Important Design Knowledge

Designing a device for simpler maintenance can significantly improve the availability and safety of a device. Still, while we imagine medical device design to be complicated, in reality, it is often very small and simple design improvements that make the difference between a useless device and a useful one.

Accessibility

The critical nature of a clinic leads to the need for unique considerations of accessibility in medical device design. In the clinic, access to key equipment and support systems may be limited. The urgency in a clinical environment can also contribute to design considerations that would be quite different from those for equipment in nonclinical operations.

Let's consider a spinal surgery procedure in a hospital operating room. The surgical plan includes a **discectomy** with **pedicle screws** and an **allograft** to restore disc height. The surgeon will be assisted by an imaging and navigation surgical system for identification of the precise anatomic location for the procedure. The medical devices that must be prepared for this procedure include spinal implants, tools, allografts, and the navigation system.

Keeping in mind that the operating room is designated a sterile room, access is restricted, and transport in and out, especially during a procedure, should be minimized. Refer to the discussions regarding the sterilization of surgical instruments in Chapter 14. The surgical instruments will be assembled in trays—usually around seven trays in all—wrapped for transport from the central sterilization unit through the hospital. In all likelihood, any implants will be delivered separately and will have been packaged and sterilized by the manufacturer. The navigation system cannot be sterilized, so, instead, it will be covered in a clear plastic sterile surgical drape.

The allograft needs to be sterile as well. In addition, it must be kept refrigerated. Operating rooms are typically not equipped with refrigerators. Therefore, the allograft must remain in outside refrigerated storage until it is needed. It can be a challenge to identify appropriate refrigeration storage nearby. Refrigerated transport of the donor tissue also may require some special logistics, especially in hot climates such as Florida, Arizona, or Texas or in the summertime.

Unique Constraints in the Operating Room

Operating rooms are typically about 400 to 600 square feet in size; a large one would be around 800 square feet. Operating rooms in older hospitals generally will be on the smaller side, especially in urban environments where real estate is at a premium. Many operating rooms were built years before large equipment such as navigation systems was available. Space can be tight, and access to power and other hookups can be precarious and limited. Extension cords should not be used in an operating room. Cables also must be clean or sterilized. The navigation system may be rendered useless because of a cable that doesn't reach a power receptacle or an unsterilized cable.

Important Design Knowledge

Small spaces and the need for appropriate cable management are only a couple of examples of the unique constraints in an operating room. If you are designing equipment for use in an operating room, you should spend some time there to observe and understand the protocols. Equipment designers must consider even the smallest detail; a small oversight might cause even the best technology to become useless in an operating room.

Wireless Access

Because wireless communication has become so prevalent in our everyday lives, it seems almost ridiculous that we must be concerned with components such as cables and connectors. But the fact is that wireless communications are highly regulated in a clinic. Early introduction of mobile phones and wireless devices in the **intensive care unit** (**ICU**) were reported to have caused catastrophic medical equipment failures, resulting in the banning of these devices in many parts of clinics (Gladman and Lapinsky 2008).

As research into **electromagnetic interference** (**EMI**) advances, the risks of EMI have been better defined, and some restrictions have been lifted. Still, wireless communications in clinics are vigilantly monitored. The Food and Drug Administration (FDA) sets standards for the use of radiofrequency (RF) devices such as Bluetooth and WiFi devices and cellular phones in coordination with the Federal Communications Commission (Day and Kerr 2014).

Important Design Knowledge

Additionally, concerns for equipment reliability and patient security are paramount. The gravity of these restrictions and concerns must be evaluated when considering wireless technology in a medical device design.

Quick Access to Equipment in the Emergency Room

Sometimes a piece of equipment may be accessible but just not quickly enough. In life or death situations, seconds count. If a doctor has to wait for a device to be set up or for a machine to boot up, the patient can die while waiting.

It might be easy to equip an emergency room well with small, inexpensive equipment, placing medical devices or machines within easy access to all patients. There may be one piece of equipment in each room. Or the emergency room may have several devices such as wheelchairs or stretchers that are stored near a group of rooms and are easily accessible. But larger and more expensive equipment may be shared more broadly within an emergency room.

Portable Equipment: Ultrasound

For example, a hospital may have only one ultrasound machine to serve the entire emergency department. An emergency room doctor may want to use the ultrasound machine to perform a **focused assessment with sonography for trauma (FAST) examination** to determine whether a patient is bleeding internally. Patients who are bleeding within the abdomen or chest must be routed to surgery immediately, and every second counts. In fact, trauma surgeons and emergency doctors speak of the **golden hour**, referring to the first hour (approximately) after the trauma occurred. This is the period in which it is much more likely that prompt medical and surgical treatment will prevent death.

In this case, the device must be truly portable. Machines and devices on bulky carts might be difficult to maneuver through the hospital or around the patient's bedside. Or they might be stored at some distance so that they are out of the way. Such equipment may be hard to reach in time for use, or it may simply be difficult to locate. Emergency rooms rarely have dedicated closets for equipment storage. Even if they do, many emergency rooms can get busy or overloaded and will focus on patient care at the expense of returning equipment to the designated storage locations.

Some equipment may be almost too portable. For example, a Sonosite portable ultrasound device is smaller and lighter than a typical school backpack and meets all the portability needs. But it is so easy to move and store that it is hard to locate quickly because it could be just about anywhere. Ideally, there should be several portable ultrasound devices throughout the emergency room, but the cost is prohibitive. Still, if a device designer can figure out how to make portable ultrasound units that are less expensive in the future, having several ultrasound machines throughout one emergency room might become very realistic.

Important Design Knowledge

While much technology has proven to be exceptionally beneficial to medical care, it must perform well under the conditions in which clinicians use it. Doctors and other clinicians are trained to work with their brains and their hands; they can care for patients in very austere environments. Healthcare professionals will quickly abandon the use of any technology that is not readily accessible, does not perform accurately, or poses any other risk or perceived risk to their patients.

Patient Adherence

Patients can be notoriously **nonadherent**. How many of you know someone who has disobeyed his or her doctor's orders by smoking or eating and drinking things they've been told to avoid? In some disease conditions, more than 40 percent of patients encounter significant risks by not adhering to the advice of their healthcare providers. It has been estimated that the annual economic burden in the United States for nonadherence is in the hundreds of millions of dollars. Additionally, the burden includes a multitude of poor health outcomes and accounts for approximately 125,000 deaths annually (Walters-Salas 2012).

Despite appearances, patients who do not adhere to their recommended regimens are not trying to be rebellious or careless. Most often it is too challenging for them to comply with their doctors' instructions. Either the regimen is too complicated or it affects the patient's lifestyle. This is compounded when someone is injured or ill and doesn't feel well enough—either physically or mentally—to perform optimally. Depression in particular is a known factor in noncompliance; a depressed patient is nearly 30 percent more likely to not follow a recommended treatment regimen (Walters-Salas 2012).

Continuous Positive Airway Pressure Adherence

Up to 10 percent of the general population suffers from **obstructive sleep apnea** (**OSA**). Because OSA interrupts normal sleep, it can be a risk factor for several diseases and increased mortality rates. It is known to decrease economic outcomes and is directly related to automobile accidents (Rotenberg, Murariu, and Pang 2016).

Treatment for many OSA patients includes the recommendation to use a **continuous positive airway pressure** (**CPAP**) device. This machine creates continuous positive airway pressure to reduce the apnea events, allowing for more normal sleep. The aim is to reduce the health risks associated with OSA. However, CPAP adherence rates range from only 30 to 60 percent. This is despite improvements over the past 20 years, such as improved machine design, quieter pumps, softer masks, and improved portability. Patients who abandon the use of CPAP cite reasons of comfort, convenience, claustrophobia, and cost. Compliance is so low that insurers require patients to undergo a trial usage before providing coverage in many cases (Rotenberg, Murariu, and Pang 2016).

Important Design Knowledge

CPAP machines represent a technology that has met FDA guidelines and has demonstrated its effectiveness. For many potential users, however, CPAP machines present too many challenges. It is not enough to introduce an effective technology. Factors that contribute to patient adherence must always be considered in a technology if the desired outcomes are to be reached. There are many opportunities for device designers to improve elements of existing technology to improve patient adherence.

Lifestyle

Brachytherapy for the treatment of breast cancer was reviewed in Chapter 14. This is also a common treatment option for prostate cancer. Radioactive seeds are implanted in the prostate tissue surrounding the cancer. If a man with prostate cancer undergoes brachytherapy treatment with seeds, the radiation dosage during the first several weeks is strong enough to affect anyone who comes into close contact with the patient. The patient may be restricted from contact with children for those weeks or may be advised against allowing children to sit in his lap. A 75-year-old grandfather might refuse this treatment because it would prevent him from enjoying time with his grandchildren. It would affect his lifestyle too drastically.

Side Effects

Brachytherapy and other forms of radiation treatment involve high dosages of radiation exposure. Often the treatment makes a patient weak or sick. Sometimes the effects of radiation treatment are so strong that a patient will choose not to complete the regimen. Improvements to radiation therapy targeting, for example, will reduce dosage and exposure, creating more options and opportunities for patients to endure the treatments.

Pain (No Needles for Diabetes)

Diabetes is a condition affecting nearly 10 percent of all Americans (Centers for Disease Control and Prevention [CDC] 2017). For years, those who must balance their lack of **endogenous** insulin with an **exogenous** dose would have to receive insulin injections. The pain and anxiety associated with needles have made such injections difficult to tolerate, and many patients therefore do not comply with their recommended treatment. In response to this, medical device engineers have introduced an entire field of **needle-free injection technology** (**NFIT**). This is comprised of a wide range of drug delivery systems that drive drugs through the skin using other forces such as electromagnetic radiation, shock waves, and pressure by gas or electrophoresis (Ravi et al. 2015).

Furthermore, in order to test their blood sugar levels, diabetic patients have to prick their fingers with a needle in order to collect a blood sample. Sometimes they have to do this several times a day or week, resulting in very sore fingertips on both hands. For years, engineers have been evaluating alternate technologies for obtaining blood sugar information that do not involve needles. Finally, in 2017, the FDA approved the first continuous blood sugar monitor that doesn't depend on backup finger prick tests. The FreeStyle Libre Flash Glucose Monitoring System attaches a small sensor to the patient's upper arm. The patient can wave a reading device across the sensor to obtain the current blood sugar level. The sensor's glucose values are based on interstitial fluid glucose levels and can be different from blood glucose levels (fingersticks), particularly during times when blood glucose is changing rapidly. But the system has smart technology to determine when blood glucose should be measured, and it identifies this in the reader.

Needle-free technologies should help to increase patient adherence and achieve good therapeutic outcomes. Any technology that can reduce pain and anxiety has the potential to significantly improve patient adherence and lead to improved health outcomes. How many people do you know who would comply better with their treatments if the pain and anxiety could be removed from them?

Cost, Access, and Ease of Use

Imagine, today, the opportunity for a person who has lost a part of a limb. Prosthesis technology is advancing rapidly, and there is technology that enables control of a prosthesis directly via wires to the brain! I can't help but recall that just 50 years ago, the best technology that my kindergarten friend could have was a wooden leg! Still, most of these opportunities are only available to a few folks in research or related situations; access is extremely limited. The cost of these technologies is often quite high. While some insurance and research subsidies exist, the prices of these prostheses are prohibitive to many in the general population who need them. Even when someone is fortunate enough to receive a high-tech prosthetic device, the device requires a lot of arduous training to achieve only simplistic function. The unfortunate result is that many who try these devices wind up abandoning them for more ordinary ones.

Important Design Knowledge

High-cost, complicated procedures and low level of benefit both lead to patient nonadherence. Device designers should always aim to enrich the performance of a device, reduce cost, and make it easier to use.

Video Games to Improve Outcomes

One way to potentially improve patient adherence and outcomes may be to incorporate video gaming into the instruction and use of a device. A recent study reviewed the work that has been done to date in this area. The study found that video games improved 69 percent of psychological therapy outcomes, 59 percent of physical therapy outcomes, 50 percent of physical activity outcomes, 46 percent of clinician skills outcomes, 42 percent of health education outcomes, 42 percent of pain distraction outcomes, and 37 percent of disease self-management outcomes (Primack et al. 2012). The researchers note that the quality of these studies was poor in that most were short-term studies and only 11 percent blinded their researchers.

Important Design Knowledge

Still these results identify an immense opportunity to include games in medical devices to improve patient outcomes. While it might be hard to immediately imagine how video gaming might be applied to a CPAP machine, it is easy to envision how much "more fun" it would be for an amputee to learn how to control a new prosthesis with the aid of a video game trainer.

Summary

The clinical environment is very unique, and unless a medical device designer considers the way a device may be used in the clinic, that device may not perform optimally or may fail. Or worse, as seen in the case of the MRI machine that resulted in the death of a six-year-old boy, it may be used improperly, leading to catastrophic results. Complex maintenance, instructions, and technology are more likely to lead to errors. Accessibility is much more severely constrained within the clinic compared with outside. Privacy, urgency, a sterile environment, and cost all contribute to these constraints. Use case evaluations must thoroughly consider how a clinic operates and how a device might be used improperly, in addition to the manufacturer's intended use.

A device also must be designed so that patients will use it effectively. Patients can be very nonadherent. Factors such as comfort, convenience, lifestyle, pain, side effects, cost, and accessibility can affect whether a patient will use a device properly or at all. These concerns invite tremendous opportunities to improve designs to better accommodate patient use scenarios. A comprehensive evaluation of clinical and patient use must be included in a medical device design strategy.

Study Questions

1. Considering that around 10 percent of the population suffers from diabetes and there are around 330 million people in the United States, how many people could conceivably benefit from more effective blood glucose monitoring systems?

2. Why has the process of getting an x-ray taken and interpreted become so much faster than it was 50 years ago? What challenges had to be overcome to allow for this increase in speed?

3. What were some of the problems with the initial use of wireless technology in clinical settings? What issues still exist?

4. Why do many patients with the good fortune to receive high-tech prosthetics choose to go back to simpler versions?

Thought Questions

1. List some use case scenarios for the different medical device examples in this chapter that you found surprising.

2. What are some strategies for addressing patient nonadherence?

3. Your organization has an idea for a revolutionary, albeit expensive, device to update how patient blood oxygen saturation is monitored in a clinical setting. What potential issues surrounding compatibility would you be worried about? How could you compensate for those issues?

4. Currently, there are ultrasound wands that can be attached to a user's cell phone and allow for cheap, portable ultrasound technology. What advantages and disadvantages can you imagine for this technology?

5. Review the last 40 years of medical record keeping in the United States. How has technology changed the way this system works? What problems remain to be solved to make it more effective?

References

Baylón, Karen, Perla Rodríguez-Camarillo, Alex Elías-Zúñiga, et al. 2017. "Past, Present, and Future of Surgical Meshes: A Review." *Membranes* 7(3):1–23. https://doi.org/10.3390/membranes7030047.

Brown, C. N., and J. G. Finch. 2010. "Which Mesh for Hernia Repair?" *Annals of the Royal College of Surgeons of England* 92(4):272–278. https://doi.org/10.1308/003588410X12664192076296.

Day, K., and P. Kerr. 2014. "FDASIA Health IT Report." U.S. Food and Drug Administration, Washington, DC, April. www.fda.gov/media/87886/download.

Gladman, Aviv S., and Stephen E. Lapinsky. 2008. "Wireless Technology in the ICU: Boon or Ban?" *Critical Care* 11(5):5–6. https://doi.org/10.1186/cc6112.

Loughlin, Sean, and Jill Schablig Williams. 2011. "Top 10 Medical Device Challenges." *Biomedical Instrumentation and Technology* 45(2):98–104. https://meridian.allenpress.com/bit/issue/45/2.

Mangan, Dan. 2001. "Child Is Killed in Hosp MRI Horror." *New York Post*, July 31, 2001. https://nypost.com/2001/07/31/child-is-killed-in-hosp-mri-horror/.

Primack, Brian A., Mary V. Carroll, Megan McNamara, et al. 2012. "Role of Video Games in Improving Health-Related Outcomes: A Systematic Review." *American Journal of Preventive Medicine* 42(6):630–638. https://doi.org/10.1016/j.amepre.2012.02.023.

Ravi, AnshDev, D. Sadhna, D. Nagpaal, and L. Chawla. 2015. "Needle-Free Injection Technology: A Complete Insight." *International Journal of Pharmaceutical Investigation* 5(4):192. https://doi.org/10.4103/2230-973x.167662.·

Rotenberg, Brian W., Dorian Murariu, and Kenny P. Pang. 2016. "Trends in CPAP Adherence Over Twenty Years of Data Collection: A Flattened Curve." *Journal of Otolaryngology Head and Neck Surgery* 45(1):1–9. https://doi.org/10.1186/s40463-016-0156-0.

U.S. Centers for Disease Control and Prevention (CDC). 2017. "National Diabetes Statistics Report, 2017." *Journal of Medical Internet Research* 20(3):e93. https://doi.org/10.2196/jmir.9515.

Walters-Salas, Trish. 2012. "The Challenge of Patient Adherence." *Bariatric Nursing and Surgical Patient Care* 7(4):186. https://doi.org/10.1089/bar.2012.9960.

CHAPTER 16

The Landscape for Medical Devices in the Twenty-First Century

Learning Objectives

Recognize current and future medical needs and goals in device design: safety, risk, efficacy, cost, accessibility, and patient adherence.

Recognize how prevention and behavior modification can contribute to improved outcomes and how biomedical engineers might offer engineering controls, tools for community support, or improved early screening, diagnosis, or genetic prediction methods.

Appreciate the implications of increasing Food and Drug Administration (FDA) regulations.

Learn about opportunities and challenges that may develop from the relatively recently introduced internet of medical things, including electronic health records and patient data management.

Learn about the use of artificial intelligence and other emerging fields in medicine.

New Terms

ergonomics

Occupational Safety and Health Administration (OSHA)

engineering controls

diabesity

medical tourism

internet of medical things (IoMT)

implantable loop recorders (ILRs)

> electronic health records (EHRs)
> Health Information Technology for Economic and Clinical Health Act (HITECH)
> evidence-based medicine
> big data
> pandemic
> personal protective equipment (PPE)

Goals for Improving Outcomes

If you are a student or engineer who is just entering the field of medical device design, you might feel slightly overwhelmed by all the technology that is already in place. Sometimes it seems like all the great ideas have already been brought to life. You might wish you had been born a hundred years ago, as discussed in Chapter 1, because there would have been so many more opportunities to develop new ideas and designs. What is left for you to contribute?

While it's true that medical devices came of age in the last century, they have a long way to go. There are still plenty of opportunities for new medical device designers to introduce disruptive innovations. More important, however, there is a tremendous need to improve on devices that have already been developed.

Safety, Risks, and Efficacy

This book has presented many examples of how medical technology has played a significant role in saving lives and improving patient outcomes. Still, we have seen that safety can be improved and risks can be reduced. In particular, the risk of infection must be eliminated. Nobody wants to go into the hospital for a routine procedure and wind up with complications due to sepsis.

We are fortunate to be living 100 years later, in a period where a patient can be cured of illnesses such as tuberculosis and even some cancers. But there are very few treatments that are 100 percent effective for 100 percent of the population. Even when a treatment is effective, it may involve serious side effects or pain. A treatment could save a life but cause harm to vital organs. A treatment may only be effective for a healthy segment of the adult population but too aggressive for older people or children to use.

There is a tremendous need to improve the efficacy of virtually every medical treatment available today. Better medical devices will play a very important role in improving healthcare outcomes and will lead to more effective treatment for a broader group of patients. The same is true if side effects or adverse events can be further reduced.

Cost, Accessibility

Many people are forced to go without sufficient healthcare because of the cost. Anything that can be done to reduce the cost of a technology likely would expand the opportunity for healthcare to more people. Technologies that cost less may be made more readily accessible to lower-income populations and third-world and austere environments around the world and may have a significant impact on global health outcomes. If you could make any of our current medical imaging technologies for half the cost, you could, for example, save an otherwise lost pregnancy or prevent a death by identifying an appendix rupture.

Even U.S. citizens who are insured will benefit from reduced costs. Healthcare spending is growing more rapidly than the U.S. gross domestic product (GDP), and the health share of GDP is expected to rise from 17.9 percent in 2017 to 19.4 percent by 2027 (Centers for Medicare and Medicaid Services [CMS] 2015). All stakeholders are seeking cost-containment solutions.

Numerous Americans receive part or all of their health insurance coverage through their employer during their years of active work and even through retirement. But the growth in healthcare spending has led employers to reduce, eliminate, or change this coverage. Americans are paying more out of pocket or skipping doctor visits.

Expensive equipment and devices have led hospitals to shop around for alternative quality equipment and device supply companies, which has negatively impacted the current business models of manufacturers. Recent studies indicate that approximately 30 percent of U.S. healthcare spending is considered waste (Shrank, Rogstad, and Parekh 2019). As strategies are developed to reduce waste and otherwise contain costs, you can be sure that pressure will increase to reduce device and equipment costs.

Tackling Adherence

Medical devices should be easier to maintain and obtain. Some devices are not efficient or perform only at a simplistic level. Some technologies are effective but cumbersome or difficult for patients to use. Devices that are easier to use or that fit better into a patient's lifestyle will improve patient adherence to treatment protocols and lead to improved outcomes. Patient adherence can be enriched by removing the tedium or pain from regimens, such as making a game out of a rehabilitation protocol.

Prevention and Behavior Modification

The onus for improving outcomes is not entirely on the medical community. Patients can assume more responsibility for their own health. This generally requires that they modify their behavior in some manner. It also requires that they are properly educated and that they have access to assistive tools and community support.

One of the best ways to improve a health outcome is to prevent a health challenge. Let's look at workplace injuries. If you were working in a factory or on a farm 100 years ago and you had to lift something heavy, you might have hurt your back, shoulders, or knee. If the injury was permanent, you would perhaps be moved to a lighter-duty (and lower-wage) job. Or you might have just been unable to work anymore.

In the middle of the twentieth century, our society began to recognize the value of the injured worker and began making efforts to prevent an injury from occurring. First, we had to learn the causes of injuries. Then we had to figure out how to avoid them. Then we had to implement techniques and strategies on the job that effectively avoided the occurrence of those injuries. This had to be a collaborative effort among medical personnel, employers, and the workers. Engineers were engaged to help evaluate hazardous movement, that is, study the **ergonomics**, and to develop devices and procedures to eliminate hazardous conditions.

This may sound fairly straightforward, but it is not. Consider that same lift injury today. You have probably seen workers who perform heavy lifts wearing back belts. Yet back belts are not recognized by the **Occupational Safety and Health Administration (OSHA)** as effective engineering controls to prevent back injury. Their effectiveness for the prevention of low back injuries has not been proven in the work environment (Reniker 2005).

Perhaps you can engineer a better device—something for which the effectiveness can be proven and not ambiguous like that of a back belt. However, OSHA's preferred approach to prevention of injuries and illnesses, including back injuries, is to use **engineering controls** to eliminate the hazardous conditions in the workplace. Engineering controls for heavy lifting on the job might include mechanical assists, adjustment of the height of the surface from which or to which material is lifted, or elimination of unnecessary bending or twisting in the task (Miles 1998).

In any case, it is important that employees and their supervisors be trained in proper lifting techniques. It is also critical that the employer institute a comprehensive ergonomics program. With proper training and support, an individual can improve his or her own health outcome tremendously by preventing injury.

While this may seem like a no brainer, ironically, half the lost workdays for registered nurses in 2016 were due to lift injuries (Dressner and Kissinger 2018). Nurses are perhaps more knowledgeable than average employees regarding injury prevention. Still, sometimes there is no good solution for lifting a patient quickly at a critical moment, so in caring for the patient, they injure themselves. Clearly, there is a need for improved engineering solutions to help prevent injuries. Maybe you can develop a device that would help nurses lift patients without exposing themselves to potential injury during critical patient transport situations.

Diabesity

We've been speaking of modifying behavior to prevent injury. Behavior modification can also prevent disease. Again, this requires education and community support, as well as efforts by individuals and their doctors. But, as discussed with regard to treatments and therapies, it can be challenging for a patient to adhere to a prescribed behavior.

Let's look at diabetes associated with obesity, aka **diabesity**. There are more than 300 million people in the world with diabetes today, and its burden on the world economy is expected to reach $490 billion in 2030 (Farag and Gaballa 2011). Why is diabesity such a rising epidemic? On face value, modifying behavior with disciplined diet and exercise may seem like a simple solution. However, we all know that this kind of self-discipline is challenging. There are multimillion-dollar industries aimed at making diets easier and exercise more fun, although long-term successful results are limited to a fraction of those who make an attempt (Wing and Phelan 2005). The reality is that there are a number of socioeconomic factors, as well as physiologic and psychological factors, that contribute to the challenge, making diabesity a very complex issue for most who have the disease. How can you envision developing a technology to help reduce the prevalence of diabesity?

Screening and Diagnosis

As mentioned, there are more than 30 million adults in the United States with diabetes. In addition, 84 million Americans have prediabetes—this equates to a prevalence of one in three adults. Prediabetes is a serious health condition in which blood sugar levels are higher than normal but not high enough yet to be diagnosed as diabetes. What is perhaps most threatening is that of those 84 million people with prediabetes, 9 of 10 of them don't even know they have the disorder (Centers for Disease Control and Prevention [CDC] 2020).

Currently, the best way to contain this epidemic is to screen for early detection. Early management of obesity and prediabetes, especially in younger individuals, can help prevent the development of diabetes and associated comorbidities (Farag and Gaballa 2011). There is an enormous need for more widespread screening and technologies that facilitate this effort.

Genetic Predictors and Targets

One area that has great potential for the diagnosis and treatment of diabetes is that of genetic studies and drug targets. Currently, genetic testing for the prediction of diabetes in high-risk individuals is of little value in clinical practice. Genetic testing for diabetes is known to have many limitations. Generally, these limitations can be described as needing risk model improvements, namely models that better correlate genetic factors with disease risk (Lyssenko and Laakso 2013). Models that better correlate genetic factors with the actual likelihood of getting the disease are needed. There is most certainly untapped potential for improved genetic studies, novel drugs, and more effective strategies for treating and preventing diabetes, obesity, and many other diseases. New medical devices and technologies are sorely needed to help advance genetic studies so that they can be clinically useful for disease screening and treatment.

Biomedical Engineering in the Twenty-First Century

We have seen how scientists of the early twentieth century worked in multidisciplinary environments to invent medical devices. Willem Einthoven was a physicist who studied the heart, Adolf Fick was a physiologist who studied partial pressure, and so on. Nearly all the scientists and engineers who have developed medical devices and equipment worked in multidisciplinary environments. Many collaborated with others whose expertise was different but complemented theirs in the context of the medical problem they were addressing.

Biomedical engineering is now recognized independently as a discipline, but it must still be viewed as a collaboration of disciplines. Foundational sciences, biology, and physics are deep and complex. It is not possible for one person to know all the science and engineering needed to address a medical challenge today. Biomedical engineers must be able to build new bridges among experts and their fields—not just across science and technology—but also into clinical practice and throughout commercial spaces. This effort will usually engage many individuals and institutions. Biomedical engineers must be able to lead teams of bright people with very different backgrounds, roles, and expertise. They must also develop an accurate, comprehensive, and up-to-date understanding of the unmet medical need and, perhaps more important, a thorough grasp of why the need has not yet been met. The more effective a biomedical engineer is in fostering these teams and bridges, the more likely it is that the team will advance the current body of knowledge and arrive at an elegant solution.

Increased Regulation

While we have seen that increased regulation is having a positive impact on medical device safety and efficacy, there are also some unintended consequences. More than 80 percent of medical devices today are introduced to the market via a 510(k) process. As you now know, this means that such devices have been deemed substantially equivalent to a predicate device—one that is already on the market. While this generally provides assurances of the device's safety, what does this say for innovation?

We have seen that the FDA has gained necessary control over many of the risk factors involved in the introduction and maintenance of medical devices in the United States. It will be important to strike a balance between safety and innovation in the future. Even the FDA recognized this challenge and published a white paper in 2004 entitled, "Stagnation or Innovation" (U.S. Food and Drug Administration [FDA] 2004).

Earlier, you learned that what is most needed to evaluate a new device is long-range clinical data. Preparing and implementing clinical trials to produce such data are also very expensive and take a long time. In a competitive marketplace, many medical device designers find themselves asking these critical questions: Can we afford to bring a brand-new idea to market? Is it worth it?

To streamline costs and time to market, many device manufacturers have implemented strategies such as overseas clinical testing and market introduction. This means that new medical technology may be available to people outside the United States before it is available within the United States. Of course, this availability is at a higher risk than our FDA deems tolerable. But those who are eager or desperate for treatment may be willing to take such risks and travel in order to receive the latest treatment. For example, by 2009, American actress Farrah Fawcett had traveled to Germany six times to receive cancer treatment. That treatment was not approved in the United States.

In fact, **medical tourism** has evolved. It was once for those living in less-developed countries seeking treatment unavailable at home who would travel to major medical centers in highly developed countries for treatment. Now the lack of access to prohibited or unavailable technology is a major incentive for people to travel outside the United States to obtain treatment. A common example is certain experimental fertility procedures.

The main reason people travel is for cost reduction. For example, surgery prices are from 30 to 70 percent lower in countries that are promoting medical tourism compared with the United States. A benefit includes staying and visiting the area after treatment.

If regulations continue to increase costs and restrict access or delay time to market, medical device designers will certainly need to consider strategies for development that are outside the United States. Increased regulation might also encourage more foreign competition. In order to compete and remain innovative at home, it will be the obligation of device developers to find ways to test new-product designs more efficiently and to produce data that will ensure low, tolerable risk.

Internet of Medical Things

You may have heard about the internet of things. The **internet of medical things** (**IoMT**) is the application of the internet to medicine and health, including data collection and analysis for research and for patient monitoring. This field is also sometimes referred to as the internet of health things. As we delve into the age of big data and the industrial internet, new opportunities and questions arise for medical technology.

Cardiac Rhythm Monitoring

The internet brings relatively new opportunities for patient monitoring and communication that will surely be expanded in the next several years. Let's look at what benefits the internet has brought to cardiac rhythm monitoring, for example. In the past century, we have seen marvelous advances in technology for cardiac monitoring. Recall Einthoven's equipment for monitoring electrocardiogram (ECG) signals that, in the early 1900s, filled up an entire room (see Chapter 2). By the time President Eisenhower had his first heart attack in 1955, the system was small enough to be transported to the president's home. Still, these devices required that the patient lay still for some duration in order to collect the data to be subsequently analyzed by the physician.

The mid-1960s brought the introduction of the first Holter monitor, a device that captured a patient's data while the patient was ambulatory and went about his or her daily activities. The earliest development of this device included an effort to transmit the signal to a physician. The first broadcast of an ECG required 85 pounds of equipment, which Holter wore on his back, and transmitted the signal approximately one block away (Figure 16.1). While effective, this device was not very practical (Roberts and Silver 1983).

With the emergence of transistor technology, the concept of telemetry was abandoned for a self-contained data storage system. Transistors allowed for the device to be made small enough to be worn (or carried on a strap). The components fit into a box about the size of a small pack, weighing just over two pounds (Roberts and Silver 1983). The patient's heart rhythm information was recorded and stored in the device. The device applied algorithms to the data in order to identify certain abnormalities. The patient would wear the monitor for a few days and then return to the doctor's office, where the doctor would read the data that were stored in the Holter monitor.

FIGURE 16.1   A Holter monitor was the first device to capture patient data while the patient was fully ambulatory.

Today, devices known as **implantable loop recorders (ILRs)** can accomplish the same task. They are about the size of a USB stick and can be implanted just under the chest skin. The patient simply has to wave a reader over the recorder, and information on the loop can be transmitted over the internet to a doctor.

The IoMT has introduced opportunities for patients and doctors to connect remotely. If desired, patient data can be monitored in real time. Doctors and medical staff can directly collect and store data from patients and save the patients' records in a database. The convenience is astounding compared with the previous century. In-home and portable monitoring is as easy as wearing a watch. Emergency notifications can be communicated based on algorithms that are applied to the data. Smart technologies are already incorporated into devices for monitoring and treating diabetes, cancer, and chronic obstructive pulmonary disease (COPD), to name a few (Sentance 2021). Opportunities for improved outcomes and healthcare cost reductions by using the internet for medical communications are in their infancy.

It is great that a patient with an ILR can transmit a daily record of his or her cardiac activity to his or her doctor via computer. But that information needs to be transmitted securely. You may have seen a spy thriller where a person who was hiding was discovered by some hacker who tracked the person via his or her pacemaker. Real or imagined, concerns for keeping patient information secure when it is accessible via the internet are valid. Measures to protect patient privacy are certain to evolve as our usage of IoMT for medical devices advances.

Electronic Health Records

With easy access and transmission of patient information, concerns arise regarding the security of the data that are stored. A paper file folder filled with a patient's record may be viewed as clumsy and cumbersome in this digital age, but the risk of unwarranted access to that record is minimal compared with electronically accessible information. However, **electronic health records (EHRs)** actually should be more secure than paper records. This is because in an EHR, information is captured regarding who views the record, including the time and duration of viewing, as well as the viewer's credentials for viewing that information. None—or very little—of this information is available when someone views a paper record.

In 2009, Congress passed the **Health Information Technology for Economic and Clinical Health Act (HITECH)** to "promote the adoption and meaningful use of health information technology." To address concerns about security breaches, revisions to this act significantly increased the penalty amounts the Secretary of Health and Human Services may impose for violations of the rules and encouraged prompt corrective action (National Archives 2009). While not a technical blockade to unwarranted hackers, these penalties should deter any reputable organization from using the data in an unauthorized manner.

HITECH was intended to accelerate the implementation of EHRs to provide for overall higher quality, better management, and safer care for patients. It authorized more than $2 billion in incentive programs. The ability to share health records electronically allows for more efficient and coordinated care for patients. This is the primary purpose of a health record. Because an EHR is quickly accessible, it offers a better chance for information to be complete and up to date at the patient's point of care. This is expected to lead to more effective diagnoses, fewer medical errors, and more reliable prescribing. An EHR system should also help providers to meet their business goals by reducing costs and improving efficiency, especially because billing and coding will be more streamlined and accurate (HealthIT 2019).

Currently, most larger institutions have completed some degree of transition to EHR systems. As you can imagine, this is a huge effort, requiring a lot of coordination among clinicians, administrators, and engineers. It will take several more years to complete this process and to more fully appreciate its benefits. The implication of EHRs for the development of medical devices and technology are yet apparent, but you can be certain that EHRs will help to identify opportunities to further improve outcomes.

Secondary Benefits

In addition to the primary benefits of EHRs, which are expected to positively impact an individual's care and an institute's performance, broader secondary benefits are anticipated.

The health information from many individuals may be combined to perform research on an aggregate population or to analyze health trends and public health issues. The data may be applied to develop new products and improve existing devices. These and other secondary benefits have existed long before EHRs, but the ability to accumulate data from paper records was cumbersome and therefore often not feasible. Data related to the performance of medical equipment, when available, have been mostly qualitative. Many uses and applications of electronic health data have not yet been imagined, just as it would have been nearly impossible for the Ringling Brothers Circus creators to imagine a Disney theme park without the advances in animation. In what ways can you envision using the data available in EHRs to reduce the risks or costs associated with a medical device?

Artificial Intelligence for Medical Decisions

It will not be long before we find new ways to take advantage of easy access to health and medical data. The ability to access and analyze data so readily is fueling the field of artificial intelligence (AI). You've already seen how a patient's data can be incorporated into cardiac monitors or smart glucose monitoring/insulin delivery devices. Many surgical robotics technologies already use various fields of data from an EHR to enhance surgical precision.

But are we really ready for AI to make comprehensive medical or surgical decisions? We are perhaps a while away from this; we've seen from the launch of self-driving cars that AI should not yet replace human drivers. Still, many drivers feel that automated emergency braking systems have helped them to avoid collisions, so it seems that some integration of AI to assist drivers may be good.

Similarly, the integration of AI into medical decision making has the potential to improve care. Already, AI is being evaluated for use in screening patients for autism spectrum disorder (Downs et al. 2019) and to aid in timely patient discharge decisions (Safavi et al. 2019). AI may contribute effectively to a surgical decision, identification of risk factors, judgments regarding postoperative management, and even shared resource use. (Loftus et al. 2019). Further, AI may help robotic prosthetics and surgical equipment "learn" rapidly to reach precision targets that are unattainable today.

Heart Surgery

In Chapters 2 and 15, we reviewed the advances in care for heart disease. From Einthoven's triangle to portable ECGs to ILR, the care and treatment for heart disease have been paramount as technology for medical devices advances. Many of the most cutting-edge treatments and techniques were first applied to heart disease.

Fields such as interventional (minimally invasive) and robotic surgery have yielded some of the most intricate and technologically advanced equipment and surgical techniques. High-tech patient monitoring, prosthetics, and artificial organ advances have almost always been applied to the management of heart disease before these developments have been applied to other conditions. Yet heart disease remains one of the top five killers today.

Even if a surgical intervention successfully saves the heart, postoperative morbidity affects up to 36 percent of cardiac surgical patients (Sanders et al. 2017). This is important to understand not only for the patient's quality of life but also for understanding how to manage costs. Chapter 12 discussed the risks of infection, but this is, unfortunately, only one of the many complications of heart surgery. It is not uncommon for someone who has had heart surgery to experience a new neurologic deficit, kidney damage, lung injury, or insufficiency, or even endocrine challenges.

Ask a family whose loved one has suffered any of these postoperative complications whether they felt that the heart surgery was worth it. Many will tell you no. At the time the surgical decision was made, however, there was a 64 percent chance (100 percent minus 36 percent) that the loved one would fare well. A more accurate prediction of the specific risk for their loved one might have guided decisions toward a better outcome.

One hundred years ago, the chances of surviving heart disease were minimal. Today, the risk of death is less than 5 percent. But more can be done to understand and prevent the large risk of postoperative morbidity and to reduce the associated high costs of their treatment.

Data

In recent years, folks have been working to collect data that will help with treatment decisions. This is known as **evidence based medicine**. Until the early 1990s, medical decisions were based on training, tradition, and mentoring more so than data. This is not to say that data were ignored; the data simply weren't available or required a herculean effort to analyze.

As the ability to analyze data became more available, insurance providers encouraged evidence-based practices, sometimes refusing coverage of practices that did not present evidence of usefulness. In previous chapters, we learned about how the FDA values evidence and how clinical research is aimed to provide evidentiary data. Still, managing and analyzing clinical data can rapidly become complex. Just ask any biostatistician.

Supercomputers and data-collection technologies such as those found in smartwatches and ILRs have introduced an unprecedented opportunity to collect a massive volume of data, known as **big data**. The processing capacity of traditional database and software management techniques is no match for the huge volume of data and the high rates of data capture that are available today. The time is ripe for new software and methods of database management that can intelligently collect and analyze large amounts of patient or human subject data. You can be certain that new information acquired from such efforts will generate evidence that will help improve decisions in medical care.

AI will play a significant role in twenty-first-century clinical data reduction and analysis, especially for bodies of data that are huge. Expect predictive and preventive modeling practices to become more commonplace as the depth and accuracy of this technology improve. For example, as discussed earlier in this chapter, improved risk models are needed for the prediction of diabetes. Presently, genetic testing can be performed on any patient, but such tests have no clinical value because of the lack of reliable predictions. There is little to no accurate correlation between specific genetic factors and the risk of diabetes. With big data and AI, a model may be developed that can describe those correlates and then use them in the clinic.

Pandemic/Summary

During the completion of this book, the COVID-19 **pandemic** took hold. Suddenly, life was different. Medical resources once assumed to be plentiful were in short supply. Hospital workers had to reuse disposable **personal protective equipment** (**PPE**) or go without. Hospitals did not have enough lifesaving drugs or equipment. At the outbreak of the pandemic, it was estimated that the United States had about 10,000 ventilators but needed 100,000. Many people who could have been saved died because of a lack of access to needed equipment and care.

Diagnosis was cumbersome, and although kits to speed up diagnosis were developed rapidly, the number of kits couldn't approach a fraction of the need. Social distancing became paramount. Parties were canceled, and tents that normally were rented for parties were propped up on lawns of hospitals and other civic grounds for makeshift diagnostic stations.

People who thought they might have been infected with the virus drove for miles and waited for hours in their cars to be tested, causing overflows so great that they created traffic jams on the highways. Testing was done only for those who were very sick. People who did not have severe symptoms just stayed isolated. Testing was not entirely accurate. People who were discharged from the hospital and who tested negative had the virus, leaving to unknowingly infect anyone with whom they came into contact.

Demand to produce more medical equipment and effective drugs simply could not be met. Homemakers sewed protective masks with their sewing machines. College laboratories were making hand sanitizer and 3D printing face shields. Researchers began characterizing the virus by its longevity on various surfaces and mode of spread. Nearly every drug company was working on developing a treatment or a vaccine.

Everyone who was nonessential was ordered to stay home or work from home. Unemployment rates in the United States skyrocketed from less than 3 percent up to 20 percent because of the virus. The world economy was severely damaged. Unthinkable choices had to be made. The loss of human lives had to be balanced against a fragile economy both locally and on a national scale. The health and welfare of the world literally depended on the speed with which the threats of the virus could be neutralized.

With not enough medical equipment available, President Trump invoked the Defense Production Act to press automobile manufacturers into the urgent business of making ventilators, respirators, masks, and other devices that were desperately needed to save lives and protect healthcare and other essential workers. As production levels ramped up, people died; the equipment could just not be made quickly enough. A leading ventilator manufacturer, Ventec, could produce 150 ventilators per month at top speed and was ramping up to 1,000 per month through its collaboration with automobile manufacturers. Still, there would not be enough ventilators in the United States until after more than 100,000 Americans died from the virus. Costs related to COVID-19 ran into trillions of dollars in the United States alone.

Meanwhile, the demand for medical equipment was so great that it stimulated unprecedented global collaborations as well as accusations of international breeches of ethics. Closer to home, hoarders sold PPE on the black market for up to 800 percent of the normal selling price. The FBI traced and arrested them for having committed federal crimes.

The FDA streamlined the approval of experimental drugs for clinical trials. Quickly, hospitals began enrolling patients in phase 2 and phase 3 trials. Research that normally would have taken several months to implement was underway in only a matter of weeks. The limiting factor was the availability of the drug.

Most assuredly, volumes will be written about how this pandemic identified the need for changes and improvements in our healthcare practices. Imagine what it would have been like if ventilators were simpler to make and cost less, if there were safer ways to protect healthcare workers, and that techniques for drug discovery could be more rapid and methods of distribution more systematic.

Recall the FDA's mission—to protect the public health by ensuring the safety, efficacy, and security of human and veterinary drugs, biological products, and medical devices. The FDA is responsible for advancing the public health by helping to speed innovations that make medical products more effective, safer, and more affordable and by helping the public get the accurate, science-based information they need to use medical products and foods to maintain and improve their health. This too is your mission. The issues brought to light during this pandemic only serve to reinforce recurring themes of the same question that we must constantly ask ourselves: how can we improve technology so that medical care is easier, cheaper, faster, safer, more reliable and more accessible—for everyone?

What can you do as a medical device engineer? Plenty!

Study Questions

1. How much healthcare spending is currently considered waste?
2. What is one of the easiest ways to avoid a health problem? How does ergonomics and engineering relate to this?
3. Define diabesity. Outline some factors contributing to its rise.
4. What is medical tourism? How has it changed in the last 50 years?
5. Give an example for how effectively implemented EHRs could change healthcare proceedings.

Thought Questions

1. What are some of the recent benefits of increased FDA regulations on medical devices? What are some of the tradeoffs?
2. Imagine that you are a local hospital administrator looking to institute a widespread prevention program for your prediabetic patients.
    *   What kind of behavior modification plans might you try to implement? What outcome would you be hoping for?
    *   What kind of medical device or diagnosis tool might you try to use? What are the opportunities and challenges with this tool?
    *   How might you use the IoMT to help you with either of these plans? What risks would you need to be mindful of?
3. What aspects of the medical field do you think are ripe for disruption by AI? Outline what makes your example a good candidate and what potential hurdles remain.

References

Centers for Medicare and Medicaid Services (CMS). 2015. "NHE Fact Sheet, 2013." Baltimore, MD. www.cms.gov/research-statistics-data-and-systems/statistics-trends-and-reports/nationalhealthexpenddata/nhe-fact-sheet.html.

Downs, Stephen M., Nerissa S. Bauer, Chandan Saha, et al. 2019. "Effect of a Computer-Based Decision Support Intervention on Autism Spectrum Disorder Screening in Pediatric Primary Care Clinics: A Cluster Randomized Clinical Trial." *JAMA Network Open* 2(12):e1917676. https://doi.org/10.1001/jamanetworkopen.2019.17676.

Dressner, Michelle, and Samuel Kissinger. 2018. "Occupational Injuries and Illnesses Among Registered Nurses." Monthly Labor Review. U.S. Bureau of Labor Statistics, Washington, DC, November. https://doi.org/10.21916/mlr.2018.27.

Farag, Youssef M. K., and Mahmoud R. Gaballa. 2011. "Diabesity: An Overview of a Rising Epidemic." *Nephrology Dialysis Transplantation* 26(1):28–35. https://doi.org/10.1093/ndt/gfq576.

HealthIT. 2019. "What Are the Advantages of Electronic Health Records?" Washington, DC. www.healthit.gov/faq/what-are-advantages-electronic-health-records.

Loftus, Tyler J., Patrick J. Tighe, Amanda C. Filiberto, et al. 2019. "Artificial Intelligence and Surgical Decision-Making," *JAMA Surgery* 2020;155(2):148-158 1–11. https://doi.org/10.1001/jamasurg.2019.4917.

Lyssenko, Valeriya, and Markku Laakso. 2013. "Genetic Screening for the Risk of Type 2 Diabetes: Worthless or Valuable?" *Diabetes Care* 36(Suppl. 2):S120–126. https://doi.org/10.2337/dcS13-2009.

Miles, John B. 1998. "Prevention of Back Injuries and Use of Back Belts." U.S. Department of Labor, Occupational Safety and Health Administration, Washington, DC. www.osha.gov/laws-regs/standardinterpretations/1998-04-06.

National Archives. 2009. "Federal Register Rules and Regulations." Washington, DC. www.hhs.gov/sites/default/files/ocr/privacy/hipaa/administrative/enforcementrule/enfifr.pdf.

National Museum of American History. 2011. "At the Heart of the Invention: The Development of the Holter Monitor." Smithsonian Institution, Washington, DC. http://americanhistory.si.edu/blog/2011/11/at-the-heart-of-the-invention-the-development-of-the-holter-monitor-1.html.

Reniker, Gary. 2005. "Back Belts: Do They Prevent Injury?" National Institute of Occupational Safety and Health, Bethesda, MD. www.cdc.gov/niosh.

Roberts, W. C., and M. A. Silver. 1983. "Norman Jefferis Holter and Ambulatory ECG Monitoring." *American Journal of Cardiology* 52(7):903–906. https://doi.org/10.1016/0002-9149(83)90439-3.

Safavi, Kyan C., Taghi Khaniyev, Martin Copenhaver, et al. 2019. "Development and Validation of a Machine Learning Model to Aid Discharge Processes for Inpatient Surgical Care" *JAMA Network Open* 2(12):1–11. https://doi.org/10.1001/jamanetworkopen.2019.17221.

Sanders, Julie, Jackie Cooper, Michael G. Mythen, and Hugh E. Montgomery. 2017. "Predictors of Total Morbidity Burden on Days 3, 5 and 8 after Cardiac Surgery." *Perioperative Medicine* 6(1):1–12. https://doi.org/10.1186/s13741-017-0060-9.

Sentence, Rebecca. 2021. "7 Examples of How the Internet of Things Is Facilitating Healthcare." Econsultancy, London, January 19. https://econsultancy.com/blog/68878-10-examples-of-the-internet-of-things-in-healthcare/.

Shrank, William H., Teresa L. Rogstad, and Natasha Parekh. 2019. "Waste in the US Health Care System: Estimated Costs and Potential for Savings." *Journal of the American Medical Association* 322(15):1501–1509. https://doi.org/10.1001/jama.2019.13978.

U.S. Centers for Disease Control and Prevention (CDC). 2020. "Division of Diabetes Translation at a Glance." National Center for Chronic Disease Prevention and Health Promotion, Washington, DC. www.cdc.gov/chronicdisease/resources/publications/aag/diabetes.htm.

U.S. Food and Drug Administration (FDA). 2004. "White Paper: Challenge and Opportunity on the Critical Path to New Medical Technologies."

Wing, Rena R., and Suzanne Phelan. 2005. "Long-Term Weight Loss Maintenance." *American Journal of Clinical Nutrition* 82(1 Suppl.):222–225. https://doi.org/10.1093/ajcn/82.1.222S.

Index

Page numbers in italics refer to figures.